Bochéin

D ✓

mark for Xerox to get variations, history, overlooked items, good defs

A PRECIS OF MATHEMATICAL LOGIC

SYNTHESE LIBRARY

A SERIES OF MONOGRAPHS ON THE

RECENT DEVELOPMENT OF SYMBOLIC LOGIC,

SIGNIFICS, SOCIOLOGY OF LANGUAGE,

SOCIOLOGY OF SCIENCE AND OF KNOWLEDGE,

STATISTICS OF LANGUAGE

AND RELATED FIELDS

Editors:

J. M. BOCHEŃSKI

A PRECIS OF MATHEMATICAL LOGIC

Translated from the French and German editions by

Otto Bird

D. REIDEL PUBLISHING COMPANY/DORDRECHT-HOLLAND

GORDON AND BREACH/NEW YORK

SOLE DISTRIBUTORS FOR NORTH AND SOUTH AMERICA

GORDON AND BREACH

SCIENCE PUBLISHERS, INC.

150 Fifth Avenue, New York 11, N.Y.

Printed in the Netherlands by D. Reidel, Dordrecht

CONTENTS

TRANSLATOR'S PREFACE

The work of which this is an English translation appeared originally in French as *Précis de logique mathématique*. In 1954 Dr. Albert Menne brought out a revised and somewhat enlarged edition in German (*Grundriss der Logistik*, F. Schoningh, Paderborn). In making my translation I have used both editions. For the most part I have followed the original French edition, since I thought there was some advantage in keeping the work as short as possible. However, I have included the more extensive historical notes of Dr. Menne, his bibliography, and the two sections on modal logic and the syntactical categories (§ 25 and 27), which were not in the original. I have endeavored to correct the typographical errors that appeared in the original editions and have made a few additions to the bibliography.

In making the translation I have profited more than words can tell from the ever-generous help of Fr. Bocheński while he was teaching at the University of Notre Dame during 1955–56.

OTTO BIRD

Notre Dame, 1959

I

GENERAL PRINCIPLES

§ 0. INTRODUCTION

0.1. *Notion and history.* Mathematical logic, also called 'logistic', 'symbolic logic', the 'algebra of logic', and, more recently, simply 'formal logic', is the set of logical theories elaborated in the course of the last century with the aid of an artificial notation and a rigorously deductive method. Leibniz (1646–1716) is generally recognized as the first mathematical logician; but it was George Boole (1815–1864) and Augustus De Morgan (1806–1878) who first presented systems in a form like those known today. Their work was taken up and furthered by C. S. Peirce (1839–1914), Gottlob Frege (1848–1925) and Giuseppe Peano (1858–1932), and then by Alfred North Whitehead and Bertrand Russell in their monumental work, *Principia Mathematica* (1910–1913). Since then active schools of mathematical logic have arisen in numerous countries, especially in America, Germany, and Poland. Progress has been rapid and is still continuing.

0.2. *Logic and mathematics.* Mathematical logic is called 'mathematical' because of its origin, since it has been developed particularly with the aim of examining the foundations of this science. There is moreover a certain external resemblance between its formulas and those of mathematics. Certain logicians also claim that mathematics is only a part of logic, although this opinion is far from receiving general approval. However, mathematical logic does not consider either numbers or quantities as such, but any objects whatsoever.

0.3. *Applications.* Mathematical logic has been successfully applied not only to mathematics and its foundations (G. Frege, B. Russell, D. Hilbert, P. Bernays, H. Scholz, R. Carnap, L. Leśniewski, T. Skolem), but also to physics (R. Carnap, A. Dittrich, B. Russell, C. E. Shannon, A. N. Whitehead, H. Reichenbach, P. Février), to biology (J. H. Woodger, A. Tarski), to psychology (F. B. Fitch, C. G. Hempel), to law and morals

(K. Menger, U. Klug, P. Oppenheim), to economics (J. Neumann, O. Morgenstern), to practical questions (E. C. Berkeley, E. Stamm), and even to metaphysics (J. Salamucha, H. Scholz, I. M. Bocheński). Its applications to the history of logic have proven extremely fruitful (J. Łukasiewicz, H. Scholz, B. Mates, A. Becker, E. Moody, J. Salamucha, K. Dürr, Z. Jordan, P. Boehner, I. M. Bocheński, S. T. Schayer, D. Ingalls.

In particular, Łukasiewicz, Salamucha, and others, have shown by the use of the methods of mathematical logic that the modern epoch has misunderstood the true sense of numerous texts of Aristotle, of almost all the logic of the Stoics, Scholastics, and the Hindus. Applications have also been made to theology (F. Drewnoswki, J. Salamucha, I. Thomas). However, it appears that we are only at the beginning. It seems certain that the logic so far developed has been used only to a small extent and that, furthermore, a very considerable development of the existing theories is possible and is in fact in the course of being accomplished.

HISTORY: The history of formal logic is a recent science, begun primarily by J. Łukasiewicz (1921) and H. Scholz (1931). – Discussions of the relation between logic and mathematics date from Leibniz and his notion of a 'mathesis universalis'; although the problem was not posed fully until Peano. – Applications and discussion of the philosophical implications of modern logic also belong to the 20th century.

LITERATURE: For the history of logic: Scholz 1; Łukasiewicz 5; Bocheński 7, 8; Beth 3; Lewis 1; Jörgensen 1; Jörgensen 2; Jordan 1. – Logic and mathe matics; Gonseth 2; PM; Russell 3; Heyting 2; Dubislav 2. – Mathematical and traditional logic: Lewis 1; Greenwood 1; Banks; Dopp. – Introductions: Carnap 1, 8; Hilbert A; Tarski 6; Reichenbach 1. Treatises: PM (the classical work, outmoded in some respects, but still the indispensable source); Hilbert B; Quine 3; Feys 5; Scholz 5; Prior; Church 6. Bibliography: Church 1 (complete for the period from 1666–1935, continued in JSL); Church 5; Beth 4 (excellent, methodically selected); cf. also bibliographies in Quine 3 and Feys 5.

§ 1. FUNDAMENTAL EXPRESSIONS AND OPERATIONS

The aim of this chapter is to enumerate the names of the fundamental logical expressions, to explain their meaning without attempting to define them strictly, and to describe some of the fundamental operations of logic. All the present chapter bears on the *names* of expressions and not on the expressions themselves. It is for this reason a metalogical theory (cf. 2.16).

1.1. *Expression, Constant, Variable*

1.11. '*Expression*' – 'a graphic sign or group of graphic signs'.
1.12. '*Expression of the system S*' – 'an expression formed in accord with the rules of system *S*'.
1.13. '*Constant of the system S*' – 'an expression which is considered to have a definite meaning in the system *S*'.
Examples: 'Peter', 'Napoleon', 'Paris', 'this book', etc.
Explanation: In defining the constant, it is necessary to add 'of the system *S*', because an expression which is a constant in a given system (e.g. in the English language) may not be a constant in another system, since the meaning of human expressions is arbitrary or conventional. The same remark applies to the expressions 'variable' (1.14), 'name' (1.33), 'functor' (1.34), 'individual variable' (1.42), etc. For the sake of simplicity this clause is omitted in most of the definitions in this section. However, it should be constantly kept in mind in understanding the definitions.
1.14. '*Variable*' – 'an expression which has no definite meaning in the system *S*, but which serves exclusively to indicate a blank where a constant can be placed'.
1.15. '*Equiform*' – two expressions are said to be 'equiform', when they have the same graphic shape, i.e. when in ordinary language they are said to be the 'same expression'.

1.2. *Substitution, syntactical category*

1.21. '*To substitute b for a in c*' or '*a by b in c*' means: 'to form an expression *d* which is equiform with *c* in everything except that in the place corresponding to *a* in *c* there is in *d* an expression equiform with *b*'.
Example: 'To substitute 'Paul' for 'Peter' in 'Peter is smoking a pipe'' means: 'form the expression: 'Paul is smoking a pipe''.

1.22. '*Syntactical category of the system S*' – 'the set of expressions which can be substituted for one another in every expression of the system *S* such that the expression formed by this substitution is itself an expression of the system *S*'.
Example: 'Peter' and 'Paul' belong to the same syntactical category of the English language, since in substituting 'Peter' for 'Paul', or vice versa, in any expression of the English language, one obtains a new expression of the English language. This is not so, for example, for 'Peter' and 'sleeps', since in substituting 'sleeps' for 'Peter' in 'Peter is smoking a pipe', one obtains 'sleeps is smoking a pipe', which is not an expression of the English language.
1.23. '*Correct substitution of variables*' – a substitution of a variable in an expression is correct when all the equiform variables of this expression have substituted for them expressions which are (1) equiform with each other and which (2) belong to the same syntactical category as the variable.
Example: The substitution of 'Peter' for 'x' in the expression '$x = x$' is correct, if two 'Peter's' are substituted for the two 'x's' by forming the expression 'Peter = Peter'. The substitution would not be correct if substitution were made only for the first 'x' by forming the expression 'Peter = x'.

1.3. *Sentence, Name, Functor*

1.31. '*Sentence of the system S*' – 'an expression which can stand (or be asserted) by itself in the system *S*'.
1.32. '*Sentential function of the system S*' – 'an expression containing variables which becomes a sentence of the system *S* when constants are substituted for all the variables'.
Explanation: 'Peter is smoking a pipe' is a sentence; but 'x is smoking a pipe' is not a sentence and is neither true nor false. It becomes a sentence when a constant is substituted for 'x'. A sentence is a graphic sign or a group of graphic signs. What is *meant* by a sentence is called a 'proposition'.
1.33. '*Name*' – 'an expression which signifies a thing (substance)'. Examples: 'Peter', 'Paris', 'this pencil'.
1.34. '*Functor*' – 'an expression which determines another expression'. Examples: 'Beautiful', 'runs', 'loves', 'it is not true that'.

Explanation: Instead of 'functor' 'operator' is sometimes used, or simply 'predicate' (Quine). The expression 'operation', frequently used, is not to be recommended, since it may lead to confusing an operation of the mind (a psychical act) with a written symbol of an extramental reality.
1.35 '*Argument of the functor F*'-'an expression determined by the functor *F*'.
Examples: 'Sky' is the argument of 'beautiful' in 'beautiful sky'; 'Peter' of 'runs' in 'Peter runs'; 'John' and 'pipe' are arguments of 'loves' in 'John loves a pipe'; 'John is sleeping' is the argument of 'it is not true that' in the sentence, 'It is not true that John is sleeping'.

1.4. *Classification of variables and functors*

1.41. '*Sentential variable*' – 'a variable for which only a sentence or a sentential function can be substituted'.
Example: In 'if *p*, then Eve is smoking a pipe', the '*p*' is a sentential variable.
1.42. '*Individual variable*' – 'a variable for which only a name of an individual can be substituted'.
Example: In '*x* is smoking a pipe', '*x*' is an individual variable.
1.43. '*Sentential functor*' – 'a functor which can have only sentences or sentential functions for arguments'.
Example: 'If ... then' is a sentential functor.
1.44. '*Individual functor*' – 'a functor which can have only names of individuals or individual variables for arguments'.
Examples: 'Drinks', 'smokes', 'detests'.
1.45. '*N-adic functor*' (where some positive integer is substituted for *n*) – 'a functor which determines *n* arguments'.
Examples: Monadic functors: 'runs', 'is untrustworthy'; dyadic functors: 'loves', 'smokes' ('*John* loves *a pipe*', '*John* smokes *a pipe*'); triadic functors: 'gives' ('*Isidore* gives *a pipe to Boniface*'); tetradic functors: 'is situated between' ('*Holland* is situated between *Germany*, *Belgium*, and *the sea*').

1.5. *Definition*

1.51. '*x* for *y*' – ''*x*' is an abbreviation for '*y*''.
1.52. 'Definition' – 'an expression formed by substitution for the variables in '*x* for *y*''.

HISTORY: The idea of variables comes from Aristotle, and the idea of syntactical categories from Husserl. The other ideas and the systematic development of the whole is the work of contemporary metalogic (Carnap, Gödel, Leśniewski, Tarski).

LITERATURE: Tarski 2; Carnap 4; a good resumé in Quine 3; Church 6.

§ 2. RULES OF WRITING

Dealing with highly abstract and complex concepts, logic has come to use artificial symbols, since words are either lacking for its concepts or, where they do exist, they are not readily usable with exactness. Logic attaches great importance to the rules of writing. This section provides two groups of rules of this kind. The first relates to the distinction between two suppositions, and the second to the technique of writing logical expressions.

2.1. *Supposition*

2.11. '*The expression X is in formal supposition*' for: 'the expression *X* signifies something different from *X* and from all expressions equiform with *X*'.

Examples: Almost all the words of ordinary language are in formal supposition. Thus when one says 'Peter is sleeping', the word 'Peter' is taken as signifying the man, Peter.

2.12. '*The expression X is in material supposition*' for: '*X* stands as a symbol of the expression *X* and all expressions equiform with *X*'.

Examples: In ''cat' is a substantive', the word 'cat' is in material supposition, since it signifies, not the animal, cat, but the word 'cat'.

2.13. *Rule:* Expressions in material supposition must be written between inverted commas, and expressions in formal supposition without them.

Example: The expression 'the cat is drinking milk' is correct, whereas 'the 'cat' is drinking milk' is not, since by putting 'cat' between inverted commas, it is affirmed that the *word* 'cat' is drinking milk. Likewise 'cat is a substantive' is incorrect, whereas ''cat' is a substantive' is correct. – Strict application of this rule is sometimes difficult in informal language, but should be striven for.

W. V. O. Quine has proposed the employment of corners, ⌜ ⌝, in addition to inverted commas for expressions in material supposition which contain variables, such as 2.12. His reason is that, since inverted commas change an expression to the name of an expression which must be considered as a whole (commas included), substitution in such an expression is not admissable.

2.14. *Rule:* Expressions in material supposition should be symbolized by expressions which are not equiform with the expressions themselves.

2.15. '*Metatheory of T*' for: 'a theory which treats of the expressions of the theory *T*'.
2.16. '*Metalogic*' for: 'metatheory of logic'.
Examples: The set of sentences of § 1 belong to metalogic.

2.2. *The placing of functors*

2.21. *Rule for the Łukasiewicz system:* All functors are placed immediately before (to the left of) their arguments.
Examples: 'loves: Peter, pipe'; 'situated between: Holland, Belgium, Germany, sea'.
2.22. In the Łukasiewicz system no parentheses are necessary.
Examples: The mathematical expression '$a + a = 2a$' is written, according to 2.21: '$= + aa2a$'. The functor '+', being dyadic, has as arguments the first two '*a*'s'; and we then have '$+ aa = 2a$'; the functor '=' is also dyadic and has as its first argument '$+ aa$' and as its second argument '$2a$'; thus we finally obtain '$= + aa2a$'.
2.23. *Rule for the Peano-Russell system:* Dyadic functors are placed between their arguments; for more complex expressions parentheses or dots are used to avoid ambiguity.
Example: In the expression '$2 + 2 \times 3$' it is necessary to use parentheses, since without them the expression can have two different meanings: '$(2 + 2) \times 3$' and '$2 + (2 \times 3)$'.

2.3. *Parentheses*

2.31. 'Parenthesis of the first degree' for: '('and')'.
2.32. 'Parenthesis of the second degree' for: '['and']'.
2.33. 'Parenthesis of the third degree' for: '{'and'}'.
2.34. 'Convex parenthesis' for: '(', or '[' or '{'.
2.35. 'Concave parenthesis' for: ')', or ']', or '}'.
2.36. *Rule:* A functor placed before a convex parenthesis has as argument the part of the expression which extends from this parenthesis to a concave parenthesis of the same degree; a functor placed after a concave parenthesis has as argument the part of the expression which extends from a convex parenthesis of the same degree up to this parenthesis.

2.4. *Dots*

2.41. *Rule:* A parenthesis of *n*-degree can be replaced by a group of *n* dots. Two expressions next to each other are considered to be separated by a group of 0 dots.

2.42. *Rule:* Dots are placed only next to functors (to which quantifiers also belong – § 11.2), and not at the beginning or end of an expression. Example: The expression '(2 + 2) × 3' is not written as '· 2 + 2 · × · 3', but as '2 + 2 · × 3'. For reasons of symmetry it may also be written '2 + 2 · × · 3'.

2.43. 'Group of dots of the first class' for: 'group of dots which stand for the functor of conjunction (§ 3.7)'.

2.44. 'Group of dots of the second class' for: 'group of dots placed to the right of a quantifier (§ 11.21–22)'.

2.45. 'Group of dots of the third class' for: 'group of dots placed to the right or left of a functor other than that of conjunction or quantification'.

2.46. *Rule:* A functor preceded or followed by a group of *n* dots of class *m* refers to the part of the expression which extends from this group to the place where there occurs (1) an equal group of dots of the same class *m* or of a higher class, or (2) a group of more than *n* dots of a lower class.

2.47. *Rule:* Conventions for subdividing the classes of dots (2.43–5) can be established as needed.

HISTORY: The theory of supposition is very old. The remainder of this section is an acquisition of the late 19th and 20th centuries. Dots were substituted for parentheses by Peano; the system of Łukasiewicz is still more recent, since it dates from 1920.

LITERATURE: 2.1; cf. preceeding §. 2.21: since the original work of Łukasiewicz is hard to obtain, one can consult: Feys 5; Bocheński 4; PM, p. 9 ff. or any of the introductions and textbooks. Further developments: Curry 2; Turing.

II

THE LOGIC OF SENTENCES

§ 3. TRUTH FUNCTORS

This chapter contains the theory of the connections between unanalysed sentences formed by functors corresponding to the English words 'not', 'or', 'if ... then', 'and', etc. These functors are called 'truth functors', because the truth of a sentence formed with them depends exclusively on the truth and not on the meaning of their arguments.

3.1. *Truth Values*

3.11. '*Truth value*' for: '1' or '0'.
Explanation: Generally, '1' is interpreted as 'true' and '0' as false, which accounts for the current definition of value: 'the value of a sentence is its truth or its falsity' (Frege).
We will consider values as symbols (in material supposition, 2.12) and not as interpreted values.
3.12. '$p = x$' for: 'the truth value of 'p' is 'x''.
Example: '$p = 1$' is read: 'the value of 'p' is truth'.
3.13. '*F is a truth functor*' for: 'the truth value of every expressions formed from F and arguments of F depends exclusively on the value of these arguments'.
Example: 'Excludes' is a truth functor, since the truth of the sentence formed from it, namely 'p excludes q', depends only on the value of 'p' and 'q'; when 'p' and 'q' both have the value '1', the sentence 'p excludes q' is false, and in all other cases it is true, independently of the meaning of 'p' and 'q'.

3.2. *Negation*

3.21. '$\{x, y\}$', where truth values are to be substituted for 'x' and 'y', for: 'a monadic truth functor in which an argument with the value '1' gives 'x' and the value '0' gives 'y''.
Remark: This may be written: $\{x, y\}1 = x$ and $\{x, y\}0 = y$.
3.22. There are 4 monadic truth functors: '$\{1, 1\}$', '$\{1, 0\}$',

'{0, 1}', and '{0, 0}'. In general, there are 2^{2n} n-adic truth functors.
3.23. '$\sim p$ (or '$\bar{p}$') or 'Np' for: '$\{0,1\}p$'.
Explanation: Read 'not p'. This functor is called 'negation'. Placed before a true sentence, it forms a false sentence; and placed before a false sentence, it forms a true sentence. The negation of a true sentence is thus false, and the negation of a false sentence is true. This is represented in the following table:
3.24.

p	$\sim p$
1	0
0	1

3.3. *Dyadic Truth Functors*

3.31. 'x, y, z, t', where truth values are substituted for 'x', 'y', 'z', and 't' for: 'the dyadic functor such that:

$$\begin{array}{ll} \text{if } p = 1 \text{ and } q = 1, & \{x, y, z, t\}pq = x \\ \text{if } p = 1 \text{ and } q = 0, & \{x, y, z, t\}pq = y \\ \text{if } p = 0 \text{ and } q = 1, & \{x, y, z, t\}pq = z \\ \text{if } p = 0 \text{ and } q = 0, & \{x, y, z, t\}pq = t\text{'.} \end{array}$$

Or in a table thus:

p q	x, y, z, t pq
1 1	x
1 0	y
0 1	z
0 0	t

or in abbreviated form:

$\{x, y, z, t\}pq$	1	0
1	x	y
0	z	t

3.32. There are $2^{2^2} = 16$ dyadic truth functors:

p	*q*	1 *V*	2 *A*	3 *B*	4 *C*	5 *D*	6 *E*	7 *F*	8 *G*
1	1	1	1	1	1	0	1	0	0
1	0	1	1	1	0	1	0	0	1
0	1	1	1	0	1	1	0	1	0
0	0	1	0	1	1	1	1	1	1

p	*q*	*O* 16	*X* 15	*M* 14	*L* 13	*K* 12	*J* 11	*I* 10	*H* 9
1	1	0	0	0	0	1	0	1	1
1	0	0	0	0	1	0	1	1	0
0	1	0	0	1	0	0	1	0	1
0	0	0	1	0	0	0	0	0	0

3.4. *Alternation or Logical Sum*

3.41. '$p \vee q$' or '*Apq*' for: '$\{1, 1, 1, 0\}pq$'.
Explanation: In our ordinary speech the alternation corresponds to 'or' taken in the non-exclusive sense (Latin 'vel'). E.g. 'he is a priest or a religious'. Such a sentence is true when one of its arguments is true, and it is false only when both are false.

3.42.

p	*q*	$p \vee q$
1	1	1
1	0	1
0	1	1
0	0	0

$\vee$	1	0
1	1	1
0	1	0

Explanation: 3.42 resembles an addition table in arithmetic:

$$1 + 1 = 2$$
$$1 + 0 = 1$$
$$0 + 1 = 1$$
$$0 + 0 = 0$$

except for the first line, where 3.42 has '1', since there is no value higher than '1' in our system. For this reason the name 'logical sum' has been given to expressions of the type '$p \vee q$' or '*Apq*'.

3.5. *Material Implication*

3.51. '$p \supset q$' (or '$p \rightarrow q$') or '*Cpq*' for: '{1, 0, 1, 1} *pq*'.
Explanation: This functor corresponds more or less to the English 'if . . . then'.
3.52.

p	q	$p \supset q$
1	1	1
1	0	0
0	1	1
0	0	1

$\supset$	1	0
1	1	0
0	1	1

3.6. *Disjunction*

3.61. '$p \mid q$' or '*Dpq*' for: '{0, 1, 1, 1} *pq*'.
3.62.

p	q	$p \mid q$
1	1	0
1	0	1
0	1	1
0	0	1

$\mid$	1	0
1	0	1
0	1	1

Explanation: The English word most closely corresponding to the functor '|' or '*D*', also called 'Sheffer's functor', is 'either . . . or'. For example, 'he is either German or French', i.e. he cannot be both at once, although he may be neither, but English.

3.7. *Conjunction or logical product*

3.71. '$p \cdot q$' (or 'p & q') or '*Kpq*' for: '1, 0, 0, 0 *pq*'.
3.72.

p	q	$p \cdot q$
1	1	1
1	0	0
0	1	0
0	0	0

$\cdot$	1	0
1	1	0
0	0	0

Explanation: The functor '·' or 'K' corresponds to the English 'and'.
3.72. resembles the multiplication table:

$$1 \times 1 = 1$$
$$1 \times 0 = 0$$
$$0 \times 1 = 0$$
$$0 \times 0 = 0$$

Hence an expression of the type '$p \cdot q$' is called a 'logical product'.
3.73. *Rule:* When dots are used for punctuation (2.41), the functor of conjunction is replaced by the most numerous groups of dots which should preceed or follow it.
Example: 'p and (q or r)' is written: '$p \cdot q \vee r$'. For '$p \cdot q$' we write: 'pq' (group of zero dots).

3.8. *Equivalence or Bi-conditional*

3.81. '$p \equiv q$' (or '$p \sim q$') or 'Epq' for: '{1, 0, 0, 1}'.
3.82.

p	q	$p \equiv q$
1	1	1
1	0	0
0	1	0
0	0	1

$\equiv$	1	0
1	1	0
0	0	1

Explanation: The '$\equiv$' functor corresponds to the English '... if and only if ...'.

3.9. *Gonseth's Graphical Representation. Terminology.*

3.91. Make a square corresponding to the abbreviated tables 3.4–7:

	$q = 1$	$q = 0$
$p = 1$		
$p = 0$		

By filling in the small square corresponding to '1' in the abbreviated tables, the following graphic representation is obtained:

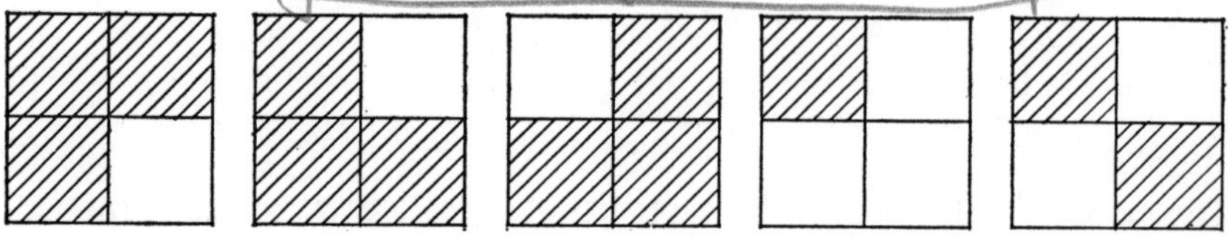

$p \vee q$ $\qquad$ $p \supset q$ $\qquad$ $p \mid q$ $\qquad$ $p \cdot q$ $\qquad$ $p \equiv q$

3.92. *Terminology of Traditional Logic*
If *Apq* ($p \vee q$), 'p' and 'q' are called 'sub-contraries'.
If *Cpq* ($p \supset q$), 'p' and 'q' are called 'subalterns'.
If *Jpq* ($p \equiv \sim q$), 'p' and 'q' are called 'contradictories'.
If *Dpq* ($p \mid q$), 'p' and 'q' are called 'contraries'.
From this we get the following *logical square:*

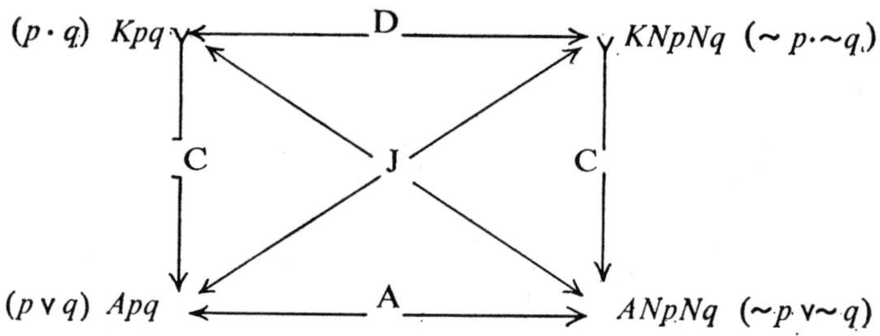

3.93. *Comparative Table of Notations:*

Definition	Peano-Russell	Łukasiewicz	Hilbert
3.23	$\sim p$	*Np*	$\bar{p}$
3.41	$p \vee q$	*Apq*	$p \vee q$
3.51	$p \supset q$	*Cpq*	$p \rightarrow q$
3.61	$p \mid q$	*Dpq*	$p \mid q$
3.71	$p \cdot q$	*Kpq*	$p \,\&\, q$
3.81	$p \equiv q$	*Epq*	$p \sim q$

Examples:

Peano-Russell:	Łukasiewicz:	Ordinary speech:
$p \cdot p \vee r$	*KpApr*	*p* and: *p* or *r*
$p \vee q \cdot q \supset r$	*KApqCqr*	*p* or *q*, and: if *q*, then *r*
$p \supset q \cdot \supset \cdot \sim q \supset \sim p$	*CCpqCNqNp*	if (if *p* then *q*), then: (if not-*q*, then not-*p*)
$p \supset q : \supset : q \supset r \cdot \supset \cdot p \supset r$	*CCpqCCqrCpr*	if: (if *p*, then *q*), then: if (if *q*, then *r*), then: (if *p*, then *r*).

The last example shows the necessity of using an artificial notation.

HISTORY: The theory was known as early as the Stoics, who gave, among others, 3.52. It was further developed by the Scholastics. It was rediscovered and further developed by Peirce in the 19th century, followed by Wittgenstein, Post and Łukasiewicz in the 20th century. The method given here is that of Feys, and the graphical representation that of Gonseth.

LITERATURE: Most of the introductions and textbooks, especially: Feys 5; Wittgenstein; Łukasiewicz T; Łukasiewicz 6. – 3.8.: Gonseth 2. On history: Łukasiewicz 5; Bocheński 2, 8; Boehner.

§ 4. EVALUATION

The problem discussed in this chapter is that of determining whether an expression is a logical law, i.e. what are the correct substitutions of its variables for it to become a true sentence. Many solutions have been given to this problem. The method of evaluation by substitution expounded here has been the most fully perfected and is the easiest.

4.1. *Definitions*

4.11. '*Logical law*' for: 'a sentential function which becomes a true sentence when constants have been correctly subtituted (1.23) for all its variables'.
4.12. '*Evaluate*' for: 'to show that an expression is or is not a law'.
4.13. '*Elementary expression*' for: 'an expression composed of '$\sim$' (or '*N*') and one truth value, or of '$\vee$', '$\supset$', '$\cdot$', '$|$', or '$\equiv$' (respectively '*A*', '*C*', '*K*', '*D*', or '*E*') and two truth values'.

4.2. *The Technique of Evaluation*

4.21. *Rule of evaluation:* (a) Determine all the possible combinations of truth values; (b) substitute the truth values of the first combination for the variables of the expression to be evaluated; (c) for the elementary expressions so obtained substitute their values according to the definitions of § 3; (d) repeat this operation until only one numeral remains; (3) if this numeral is '0', the expression is not a law, (f) if it is '1', substitute in the other combinations in the same way; and (g) if all the combinations have '1' as their result, the expression is a law.
4.22. The possible combinations of truth values are 2^n for *n* non-equiform variables.
If all variables are equiform, there are 2 values: $p = 1, p = 0$.
For 2 non-equiform variables, there are 4 values:

(1) $p = 1, \quad q = 1$
(2) $p = 1, \quad q = 0$
(3) $p = 0, \quad q = 1$
(4) $p = 0, \quad q = 0$

For 3 non-equiform variables, there are 8 values:

(1) $p = 1,\quad q = 1,\quad r = 1$
(2) $p = 1,\quad q = 1,\quad r = 0$
(3) $p = 1,\quad q = 0,\quad r = 1$
(4) $p = 1.\quad q = 0,\quad r = 0$
(5) $p = 0,\quad q = 1,\quad r = 1$
(6) $p = 0,\quad q = 1,\quad r = 0$
(7) $p = 0,\quad q = 0,\quad r = 1$
(8) $p = 0,\quad q = 0,\quad r = 0$

For 4 non-equiform variables, there are 16 values, etc.

4.23. *Truth values of the elementary expressions:* The following table facilitates the substitution of values for the elementary expressions:

Expr.	Val.	Expr.	Val.	Expr.	Val.	Expr.	Val.	Expr.	Val.
$1 \vee 1$	1	$1 \supset 1$	1	$1 \mid 1$	0	$1 \cdot 1$	1	$1 \equiv 1$	1
$1 \vee 0$	1	$1 \supset 0$	0	$1 \mid 0$	1	$1 \cdot 0$	0	$1 \equiv 0$	0
$0 \vee 1$	1	$0 \supset 1$	1	$0 \mid 1$	1	$0 \cdot 1$	0	$0 \equiv 1$	0
$0 \vee 0$	0	$0 \supset 0$	1	$0 \mid 0$	1	$0 \cdot 0$	0	$0 \equiv 0$	1
~ 1	0								
~ 0	1								
A11	1	C11	1	D11	0	K11	1	E11	1
A10	1	C10	0	D10	1	K10	0	E10	0
A01	1	C01	1	D01	1	K01	0	E01	0
A00	0	C00	1	D00	1	K00	0	E00	1
N1	0								
N0	1								

Example: Evaluate the following expression: '*CCpqCNqNp*'.
There are two non-equiform variables: '*p*' and '*q*'; hence there are 4 substitutions:

(a) $p = 1, q = 1$	(b) $p = 1, q = 0$	(c) $p = 0, q = 1$	(c) $p = 0, q = 0$
CCpqCNqNp	*CCpqCNqNp*	*CCpqCNqNp*	*CCpqCNqNp*
*CC*11*CN*1*N*1	*CC*10*CN*0*N*1	*CC*01*CN*1*N*0	*CC*00*CN*0*N*0
C 1 C 0 0	C 0 C 1 0	C 1 C 0 1	C 1 C 1 1
C 1 1	C 0 0	C 1 1	C 1 1
1	1	1	1

Since all 4 substitutions yield the truth value '1', the expression is always true and hence a logical law (4.21).

For the Peano-Russell notation the technique of evaluation is the same, except that it is necessary to omit the dots and parentheses. For this reason it is easier to evaluate expressions in the Łukasiewicz notation.

HISTORY: Cf. § 3.

LITERATURE: The clearest expositions are: Scholz 5; Quine 3; Feys 5.

§ 5. EQUIVALENCES

This chapter contains the simplest and most useful logical laws in the form of equivalences. They are given in two notations, first that of Peano-Russell, and then that of Łukasiewicz.

5.1. contains laws in which all the variables are equiform.

5.2–6. contains laws with two or three non-equiform variables grouped according to the functor of the first argument.

5.1. *Laws in which all the variables are equiform*

5.11. $p \equiv p$	*Epp*	Principle of Identity
5.12. $\sim\ \sim p \equiv p$	*ENNpp*	Principle of Double Negation
5.13. $\sim\ \sim\ \sim p \equiv\ \sim p$	*ENNNpNp*	Principle of Triple Negation
5.14. $\sim p \equiv p \mid p$	*ENpDpp*	Reduction of Negation (cf. 5.45)
5.15. $p \vee p \cdot \equiv \cdot p$	*EAppp*	1st law of Tautology
5.16. $pp \equiv p$	*EKppp*	2nd law of Tautology

5.2. *Laws of the Sum (Alternation)*

5.211. $p \vee q \cdot \equiv \cdot \sim p \supset q$	*EApqCNpq* Cf. 5.311
5.212. $p \vee q \cdot \equiv \cdot \sim p \mid \sim q$	*EApqDNpNq* Cf. 5.41 Reduction of Alternation
5.213. $p \vee q \cdot \equiv \cdot \sim \cdot \sim p \sim q$	*EApqNKNpNq* De Morgan's 3rd Law
5.214. $p \vee q : \equiv : p \supset q \cdot \supset q$	*EApqCCpqq*
5.22. $p \vee q \cdot \equiv \cdot q \vee p$	*EApqAqp* Commutative Law of the Sum
5.23. $p \cdot \vee \cdot q \vee r : \equiv : p \vee q \cdot \vee \cdot r$	*EApAqrAApqr* Associative Law of the Sum
5.24. $p \cdot \vee \cdot qr : \equiv : p \vee q \cdot p \vee r$	*EApKqrKApqApr* Distributive Law of the Sum
5.25. $p \cdot \vee \cdot p \vee q : \equiv : p \vee q$	*EApApqApq* 1st law of Simplification of the Sum
5.26. $p \cdot \vee \cdot pq : \equiv : p$	*EApKpqp* 2nd Law of Simplification of the Sum

5.27. $\sim \cdot p \vee q \cdot \equiv \cdot \sim p \cdot \sim q$ — *ENApqKNpNq*
De Morgan's 1st Law

5.3. *Laws of Implication*

5.311. $p \supset q \cdot \equiv \cdot \sim p \vee q$ — *ECpqANpq*

5.312. $p \supset q \cdot \equiv \cdot p \mid \sim q$ — *ECpqDpNq*
Reduction of Implication

5.313. $p \supset q \cdot \equiv \cdot \sim \cdot p \sim q$ — *ECpqNKpNq*

5.314. $p \supset q \cdot \equiv \cdot p \equiv pq$ — *ECpqEpKpq*

5.315. $p \supset q : \equiv : q \cdot \equiv \cdot p \vee q$ — *ECpqEqApq*

5.32. $p \supset q \cdot \equiv \cdot \sim q \supset \sim p$ — *ECpqCNqNp*
Law of Simple Contraposition

5.321. $p \supset \sim q \cdot \equiv \cdot q \supset \sim p$ — *ECpNqCqNp*
2nd Law of Simple Contraposition

5.322. $\sim p \supset q \cdot \equiv \cdot \sim q \supset p$ — *ECNpqCNqp*
3rd Law of Simple Contraposition

5.33. $p \cdot \supset \cdot q \supset r : \equiv : q \cdot \supset \cdot p \supset r$ — *ECpCqrCqCpr*
Law of Simple Commutation

5.34. $pq \supset r : \equiv : p \cdot \supset \cdot q \supset r$ — *ECKpqrCpCqr*
1st Law of Exportation

5.35. $pq \supset r : \equiv : q \cdot \supset \cdot p \supset r$ — *ECKpqrCqCpr*
2nd Law of Exportation

5.36. $p \cdot \supset \cdot p \supset q : \equiv : p \supset q$ — *ECpCpqCpq*

5.37. $pq \supset r \cdot \equiv \cdot \sim rq \supset \sim p$ — *ECKpqrCKNrqNp*
1st Law of Syllogistic Contraposition

5.38. $pq \supset r \cdot \equiv \cdot p \sim r \supset \sim q$ — *ECKpqrCKpNrNq*
2nd Law of Syllogistic Contraposition

5.39. $\sim \cdot p \supset q \cdot \equiv \cdot p \sim q$ — *ENCpqKpNq*

5.4. *Laws of Disjunction*

5.41. $p \mid q \cdot \equiv \cdot \sim p \vee \sim q$ — *EDpqANpNq*

5.42. $p \mid q \cdot \equiv \cdot p \supset \sim q$ — *EDpqCpNq*

5.43. $p \mid q \cdot \equiv \cdot \sim \cdot pq$ — *EDpqNKpq*

5.44. $p \mid q \cdot \equiv \cdot q \mid p$ — *EDpqDqp*
Commutation of Exclusion

5.45. $p \mid p \cdot \equiv \cdot \sim p$ — *EDppNp*
Reduction of Negation

5.46. $\sim \cdot p \mid q \cdot \equiv \cdot pq$ — *ENDpqKpq*

5.5. *Laws of the Product* (*Conjunction*)

5.511. $pq \cdot \equiv \cdot \sim \cdot \sim p \vee \sim q$ — *EKpqNANpNq*
De Morgan's 4th Law

5.512. $pq \cdot \equiv \cdot \sim \cdot p \supset \sim q$ — *EKpqNCpNq*

5.513. $pq \cdot \equiv \cdot \sim \cdot p \mid q$ — *EKpqNDpq*
Reduction of Conjunction

5.52. $pq \equiv qp$ — *EKpqKqp*
Commutative Law of the Product

5.53. $p \cdot qr \cdot \equiv \cdot pq \cdot r$ — *EKpKqrKKpqr*
Associative Law of the Product

5.54. $p \cdot q \vee r \cdot \equiv \cdot pq \vee pr$ — *EKpAqrAKpqKpr*
Distributive Law of the Product

5.55. $p \cdot p \vee q \cdot \equiv \cdot p$ — *EKpApqp*
1st Law of Simplifying the Product

5.56. $p \cdot pq \cdot \equiv \cdot pq$ — *EKpKpqKpq*
2nd Law of Simplifying the Product

5.57. $\sim \cdot pq \cdot \equiv \cdot \sim p \vee \sim q$ — *ENKpqANpNq*
De Morgan's 2nd Law

5.58. $\sim \cdot pq \cdot \equiv \cdot p \mid q$ — *ENKpqDpq*

5.6. *Laws of Equivalence*

5.611. $p \equiv q \cdot \equiv \cdot pq \vee \sim p \sim q$ — *EEpqAKpqKNpNq*

5.612. $p \equiv q \cdot \equiv \cdot p \supset q \cdot q \supset p$ — *EEpqKCpqCqp*

5.62. $p \equiv q \cdot \equiv \cdot q \equiv p$ — *EEpqEqp*
Commutative Law of Equivalence

5.63. $p \cdot \equiv \cdot q \equiv r : \equiv : p \equiv q \cdot \equiv \cdot r$ *EEpEqrEEpqr* Associative Law of Equivalence

5.64. $p \equiv q \cdot \equiv \cdot \sim p \equiv \sim q$ *EEpqENpNq* Inversion of Equivalence

5.65. $p \equiv q \cdot \equiv \cdot \sim q \equiv \sim p$ *EEpqENqNp* 1st Contraposition of Equivalence

5.66. $\sim p \equiv q \cdot \equiv \cdot \sim q \equiv p$ *EENpqENqp* 2nd Contraposition of Equivalence

5.67. $p \equiv \sim q \cdot \equiv \cdot q \equiv \sim p$ *EEpNqEqNp* 3rd Contraposition of Equivalence

5.7. *Rules of Transformation* by which these laws can be developed so as to yield still further laws.

5.71. 'Chief functor of *X*' for 'the largest point-group of '$\equiv$' in *X*'.
5.72. *Rule for Inversion:* If *X* is one of the laws of § 5, left column, the expression formed by substituting the part of *X* which follows the chief functor for that which preceeds it, and *vice versa,* is itself a logical law.
Example: Since (5.16) '$pp \equiv p$' is a law, '$p \equiv pp$' is also a law.
5.73. *Rule for substitution of implication:* If *X* is a law of § 5, left column, the expression formed by substituting '$\supset$' for its chief functor is also a law. Or, in Łukasiewicz notation, the expression formed by substituting '*C*' for its first '*E*' is also a law.
Example: Since (5.11) '$p \equiv p$', or '*Epp*' is a law, so also '$p \supset p$' or '*Cpp*' is a logical law.

HISTORY: Almost all the laws of this section were known to the Scholastics, including those wrongly named after De Morgan (5.27, 5.57, 5.213, 5.511).

LITERATURE: An all but complete enumeration of the laws used in practice is given in PM *2 – *5 and in the textbooks, especially Feys 5.

§ 6. 'FIRST PRINCIPLES' AND IMPLICATIONS

The laws enumerated here are, together with the equivalences of § 5, the most important. 6.1. contains three laws which, with 5.11. are known as the 'first principles' of traditional logic; they are presented in the notation of the Sentential Calculus. The arrangement is the same as that of § 5.

6.1. *'First Principles'*

6.11.	$\sim \cdot p \sim p$	$NKpNp$	Principle of Non-Contradiction
6.12.	$p \mid \sim p$	$DpNp$	
6.13.	$p \vee \sim p$	$ApNp$	Principle of Excluded Middle
6.2.	*Characteristic Laws of Implication*		
6.21.	$p \cdot \supset \cdot q \supset p$	$CpCqp$	1st Paradoxical law ('Verum sequitur ad quodlibet')
6.22.	$\sim p \cdot \supset \cdot p \supset q$	$CNpCpq$	2nd Paradoxical law ('Ex falso sequitur quodlibet')
6.23.	$p \supset \sim p \cdot \supset \cdot \sim p$	$CCpNpNp$	1st Reductio ad absurdum
6.24.	$p \cdot \supset \cdot \sim p \supset q$	$CpCNpq$	
6.25.	$p \sim p \supset \sim p$	$CKpNpNp$	2nd Reductio ad absurdum
6.26.	$p \cdot \supset \cdot p \vee q$	$CpApq$	Law of the new factor
6.27.	$pq \supset p$	$CKpqp$	1st law of the *a fortiori*
6.271.	$pq \supset q$	$CKpqq$	2nd law of the *a fortiori*
6.281.	$pq \cdot \supset \cdot p \supset q$	$CKpqCpq$	
6.282.	$p \equiv q \cdot \supset \cdot p \supset q$	$CEpqCpq$	

6.3. *Laws of the Syllogism*

6.31.	$q \supset r : \supset : p \supset q \cdot \supset \cdot p \supset r$	$CCqrCCpqCpr$
6.32.	$p \supset q : \supset : q \supset r \cdot \supset \cdot p \supset r$	$CCpqCCqrCpr$
6.33.	$p \supset q :\cdot \supset :\cdot q \supset r : \supset : r \supset s \cdot \supset \cdot p \supset s$	$CCpqCCqrCCrsCps$
6.34.	$p \supset q :: \supset :: q \supset r :\cdot \supset :\cdot r \supset s : \supset : s \supset t \cdot \supset \cdot p \supset t$	$CCpqCCqrCCrsCCstCpt$
6.35.	$p \supset q \cdot q \supset r \cdot \supset \cdot p \supset r$	$CKCpqCqrCpr$
6.36.	$p \supset q \cdot q \supset r \cdot r \supset s \cdot \supset \cdot p \supset s$	$CKKCpqCqrCrsCps$
6.37.	$p \supset q \cdot q \supset r \cdot r \supset s \cdot s \supset t \cdot \supset \cdot p \supset t$	$CKKKCpqCqrCrsCstCpt$
6.38.	$p \supset q : \supset : r \vee p \cdot \supset \cdot r \vee q$	$CCpqCArpArq$

6.4. *Modes of the Hypothetical Syllogism*

6.41. $p : \supset : p \supset q \cdot \supset \cdot q$ *CpCCpqq* Modus ponendo ponens 1°
6.42. $p \supset q \cdot p \cdot \supset \cdot q$ *CKCpqpq* Modus ponendo ponens 2°
6.421. $p \supset \sim q \cdot p \cdot \supset \cdot \sim q$ *CKCpNqpNq*
6.422. $\sim p \supset q \cdot \sim p \cdot \supset \cdot q$ *CKCNpqNpq*
6.423. $\sim p \supset \sim q \cdot \sim p \cdot \supset \cdot \sim q$ *CKCNpNqNpNq*
6.43. $\sim q : \supset : p \supset q \cdot \supset \cdot \sim p$ *CNqCCpqNp*
Modus tollendo tollens 1°
6.44. $p \supset q \cdot \sim q \cdot \supset \cdot \sim p$ *CKCpqNqNp*
Modus tollendo tollens 2°
6.441. $p \supset \sim q \cdot q \cdot \supset \cdot \sim p$ *CKCpNqqNp*
6.442. $\sim p \supset q \cdot \sim q \cdot \supset \cdot p$ *CKCNpqNqp*
6.443. $\sim p \supset \sim q \cdot q \cdot \supset \cdot p$ *CKCNpNqqp*

6.5. *Modes of the Disjunctive and Copulative Syllogism*

6.51. $\sim p : \supset : p \vee q \cdot \supset \cdot q$ *CNpCApqq*
Modus tollendo ponens 1°
6.511. $\sim q : \supset : p \vee q \cdot \supset \cdot p$ *CNqCApqp*
6.52. $p \vee q \cdot \sim p \cdot \supset \cdot q$ *CKApqNpq*
Modus tollendo ponens 2°
6.521. $p \vee q \cdot \sim q \cdot \supset \cdot p$ *CKApqNqp*
6.522. $p \vee \sim q \cdot \sim p \cdot \supset \cdot \sim q$ *CKApNqNpNq*
6.523. $\sim p \vee q \cdot p \cdot \supset \cdot q$ *CKANpqpq*
6.524. $\sim p \vee \sim q \cdot p \cdot \supset \cdot \sim q$ *CKANpNqpNq*
6.53. $p : \supset : p \mid q \cdot \supset \cdot \sim q$ *CpCDpqNq*
Modus ponendo tollens 1°
6.531. $q : \supset : p \mid q \cdot \supset \cdot \sim p$ *CqCDpqNp*
6.532. $p : \supset : \sim \cdot pq \cdot \supset \cdot \sim q$ *CpCNKpqNq*
6.54. $p \mid q \cdot p \cdot \supset \cdot \sim q$ *CKDpqpNq* Modus ponendo tollens 2°
6.541. $p \mid \sim q \cdot p \cdot \supset \cdot q$ *CKDpNqpq*
6.542. $\sim p \mid q \cdot \sim p \cdot \supset \cdot \sim q$ *CKDNpqNpNq*
6.543. $\sim p \mid \sim q \cdot \sim p \cdot \supset \cdot q$ *CKDNpNqNpq*

6.6. *Laws of Composition and Dilemmas*

6.61. $p \supset q \cdot p \supset r \cdot \supset \cdot p \supset qr$ *CKCpqCprCpKqr*
Law of multiplying the consequent

6.62.	$p \cdot \supset \cdot q \supset pq$	*CpCqKpq*
6.63.	$p \supset q \cdot r \supset s : \supset : pr \supset qs$	*CKCpqCrsCKprKqs* Multiplication of both sides
6.64.	$p \supset q \cdot r \supset s : \supset : p \vee r \cdot \supset \cdot q \vee s$	*CKCpqCrsCAprAqs* Addition of both sides
6.65.	$p \supset r \cdot q \supset r : \supset : p \vee q \cdot \supset \cdot r$	*CKCprCqrCApqr* 1st constructive dilemma
6.66.	$p \supset r \cdot \sim p \supset r \cdot \supset \cdot r$	*CKCprCNprr* 2nd constructive dilemma
6.67.	$\sim q \sim r : \cdot \supset : \cdot p \cdot \supset \cdot q \vee r : \supset : \sim p$	*CKNqNrCCpAqrNp* 1st destructive dilemma
6.671.	$p \cdot \supset \cdot qr : \sim q \sim r : \supset : \sim p$	*CKCpKqrKNqNrNp* 2nd destructive dilemma
6.68.	$p \supset q \cdot p \supset \sim q \cdot \supset \cdot \sim p$	*CKCpqCpNqNp* 3rd reductio ad absurdum

HISTORY: 6.24 is found as early as Democritus and Plato; 6.11, 6.13, 6.35, 6.44 were known to Aristotle; 6.41, 6.43, 6.52, 6.532 were posited by the Stoics as 'indemonstrable' laws. Almost all of the laws of this and the preceeding sections were reached by the Scholastics, clearly independently of the Stoics. 6.63 (along with some others) was discovered, or rather re-discovered, by Leibniz and pleased him so much that he called it the 'praeclarum theorema'.

LITERATURE: Cf. § 5.

§ 7. AXIOMATIC SYSTEM

The theory of an axiomatic system represents an ideal of deductive method which has always been taught by logic. This method was first applied in full rigor to logic itself and at present has been so fully developed that it can be applied to many other domains. This chapter presents the theory in succinct form by sacrificing rigor for clarity; a rigorous development would call for long explanations.

7.1. *Definitions*

7.11. '*Axiomatic System*' for 'the set of expressions falling into two classes such that the elements of the second are derived from the first by the application of explicitly formulated rules'.

7.12. An axiomatic system contains terms, sentences, and laws, rules of definition for terms, rules of formation for sentences, and rules of deduction for laws.

7.2. *Terms and Definitions*

7.21. '*Term of the system S*' for: 'expression of the system *S* of which no part is an expression of the system *S*'.

Example: '$\vee$' (or '*A*'), '*p*', '*q*' are terms of the Sentential Calculus, whereas '$p \vee q$' is not, since it contains parts which are sentences, namely '*p*' and '*q*'.

7.22. '*To define X by Y*' for: 'to form an expression which indicates that *X* can be substituted for *Y*'. Definition in this sense is not the determination or explanation of an essence, a concept, or a word, but only the positing that one sign can be used for another; in general it is the abbreviation of a longer series of signs.

7.23. '*Primitive term of the system S*' for: 'term of the system *S* which is not defined in the System *S*'.

7.24. '*Derived term of the system S*' for: 'term defined in system *S*'.

7.25. '*Rule of definition of the system S*' for: 'rule which indicates the correct way of defining the derived terms of the system *S*'.

7.26. *Rule:* All primitive terms and rules of definition of the axiomatic system must be stated explicitly, and all terms which are not primitive must be explicitly defined.

7.3. *Sentences and Rules of Formation*

7.31. '*Rule of formation of the system S*' for: 'rule which indicates how the terms of the system S can be formed into sentences of the system *S*'.
Example: One of the rules of the Łukasiewicz system is 'a group of terms composed of '*C*' and two variables is a sentence'. In many systems all expressions are sentences.
7.32. *General rule of formation:* All sentences in the system *S* must be formed exclusively from terms of *S* according to the manner indicated by the formation rules of *S*.

7.4. *Laws and Deduction*

7.41. '*Law of the system S*' for: 'sentence asserted in the system *S*'.
7.42. '*To deduce Y from X in the system S*' for: 'to show that if *X* is a law of *S*, the rules of *S* allow the assertion of *Y*'.
7.43. '*Axiom of the system S*' for: 'law of the system *S* which is not deduced in the system *S*'.
7.44. '*Theorem of the system S*' for: 'law of the system *S* deduced from the axioms of *S* by the application of the rules of *S*'.
7.45. '*Rule of deduction of the system S*' for: 'rule which indicates the correct way of deducing in the system *S*'.
7.46. *Rule:* All axioms and all rules of deduction of the axiomatic system must be stated explicitly; all other asserted sentences must be deduced explicitly.
7.47. *Rule:* The application of laws and definitions in the deduction of a theorem must be formulated explicitly in a special expression called the 'proof', 'probative verse', or 'derivational line'.
Example: § 8 contains several examples of this with explanations.
7.48. '*X implies Y inferentially in the system S*' for: 'the rules of deduction of the system *S* allow *Y* to be deduced from *X*'.
Explanation: Material implication must not be confused with inferential implication. The former, for example, holds among all true sentences, which is not the case, however, for the latter. The 'if' of ordinary English is closer to inferential than to material implication.

7.5. *Formalism*

7.51. '*Formalized system*' for: 'axiomatic system whose rules concern exclusively the graphical form of expressions and all of whose axioms and rules are explicitly formulated'.

7.52. *Rule:* The formalized system, once its axioms and rules are established, must be developed solely in virtue of its rules and without any reference to the semantic meaning of the expressions that are used.
7.53. The formalized axiomatic system as such has no semantic meaning and can receive diverse interpretations.
Example: The system expounded in §§ 4–6 must be considered as a set of letters which symbolize *nothing*. For example, the '*C*' must not be taken as the symbol for implication according to the ordinary meaning of 'if', but it must be taken exclusively as a functor defined according to table 3.52. In §§ 9, 12, and 16 we will give three different interpretations to it which facilitate the understanding of § 3.

7.6. *Consistency*

7.61. '*Non-contradictory system*' for: 'axiomatic system whose rules of deduction do not allow a sentence to be deduced along with the negation of this sentence'.
7.62. In a complete system which is contradictory any sentence can be deduced.
Explanation: In virtue of 6.24 a sentence asserted at the same time as its negation allows us to deduce '*q*'. By substitution we can then obtain any sentence we want. As a result the distinction between true and false sentences vanishes, and science is no longer possible. It is this that led Aristotle to say that the principle of non-contradiction (6.11) is the first principle of logic.

7.7. *Completeness and Independence*

7.71. '*Complete system in a wide sense*' for: 'axiomatic system which contains all the true sentences of a given domain'. It can also be said that no sentence of a given domain is true if it is not derivable in the system.
7.72. '*Complete system in a strict sense*' for: 'axiomatic system in which each sentence which is not a law is the negation of one of its laws'.
7.73. '*System with independent axioms*' for: 'axiomatic system in which no axiom can be deduced from other axioms of the system by the application of the rules of the system'.

7.8. *Rules*

7.81. Every axiomatic system must be formalized and non-contradictory.

7.82. One must attempt to establish complete systems in the strict sense with independent axioms.

HISTORY: The axiomatic system is one of the discoveries due to the genius of Aristotle (*Posterior Analytics*); especially cultivated by mathematicians (Euclidian system), it has received formal and rigorous elaboration from contemporary meta-logicians (cf. § 26.3).

LITERATURE: Elementary expositions: Carnap 1; Tarski 6; Hilbert A; Prior. Rigorous metalogical elaborations: Cf. § 26; also Bernays; Schröter 1; Schröter 2; Woodger 3; Carnap 5. There is an extensive literature on methods for proving consistency, completeness, and independence of axioms, of which the most notable is the work of Gödel.

§ 8. A SYSTEM OF THE LOGIC OF SENTENCES

This section provides by way of example an axiomatic system of the sentential calculus. The method employed for developing it is the most rigorous of those that are known. Only the foundations (definitions, axioms, rules, etc.) and some of the beginning demonstrations are given here.

8.1. *Primitive Terms, Rule of Definition and Rules of Formation*

8.11. *Primitive terms:* '*D*' – dyadic functor; '*p*', '*q*', '*r*', '*s*' – sentential variables.

8.12. *Rule of Definition:* A new term can be introduced into the system by formulating a group of terms, called the 'definition', and consisting of: (1) an expression which contains the new term and in which all the others are terms of the system; (2) '='; (3) an expression which contains only primitive terms or terms already defined.

8.13. *Rules of Formation:* (1) a variable is a sentence; (2) a group of terms consisting of '*N*' followed by one sentence is a sentence; (3) a group of terms consisting of '*A*', '*C*', '*D*', '*E*', or '*K*' followed by two sentences is a sentence.

8.2. *Definitions*

8.21. *Np* = *Dpp* (cf. 5.14)
8.22. *Apq* = *DNpNq* (cf. 5.213)
8.23. *Cpq* = *ANpq* (cf. 5.311)
8.24. *Kpq* = *NANpNq* (cf. 5.511)
8.25. *Epq* = *KCpqCqp* (cf. 5.612)

8.3. *Rules of Deduction*

8.31. *Rule of Substitution:* A sentence may be substituted for a variable, but the same sentence must be substituted for all equiform occurrences of variables in the expression.

8.32. *Rule of substitution by definition:* A definition may be substituted for the expression which it defines in a sentence, and reciprocally, without being substituted for all equiform occurrences of that expression.

8.33. *Rule of detachment:* If a sentence consisting of '*C*' followed by two sentences is a law of the system, and if a sentence equiform with the first of these sentences is a law of the system, a sentence equiform with the second can be posited as a law of the system.

8.4. *Axioms*

8.41. *CAppp* (cf. 5.15)
8.42. *CpApq* (cf. 6.26)
8.43. *CApqAqp* (cf. 5.22)
8.44. *CCpqCArpArq* (cf. 6.38)

8.5. *Deduction:*

$8.44\ r/Nr \times 8.23\ p/r,\ q/p \times 8.23\ p/r = 8.51$

8.51. *CCpqCCrpCrq* (cf. 6.38)

Explanation: The probative verse or derivational line of theorem 8.51 is to be read as follows: Take axiom 8.44; for '*r*' substitute '*Nr*'; apply definition 8.23 after substituting '*r*' for '*p*' and '*p*' for '*q*'; apply the same definition 8.23 again but this time substitute '*r*' for '*p*'; and one will thus obtain the theorem to be proved 8.51. Or to write it out in full, we have:

8.44. *CCpqCAr pAr q*
r/*Nr* (substitute '*Nr*' for '*r*') *CCpqCANrpANrq*
8.23 $Cpq = ANpq$
p/r $Crq = ANrq$
q/p $Crp = ANrp$

We can now put '*Crp*' for '*ANrp*' in our rewritten 8.44 and have:

CCpqC CrpANrq

8.23
p/r $Crq = ANrq$
For '*ANrq*' put '*Crq*': *CCpqC Crp Crq*, which is 8.51.

8.51. $p/App,\ q/p,\ r/p = C8.41 - C8.42\ q/p - 8.52$

8.52. *Cpp*

Explanation: After carrying out the substitutions indicated at the beginning, we obtain the expression:

CCApppCCpAppCpp

which is composed of (1) '*C*', (2) '*CAppp*' which is an expression equiform with 8.41, (3) '*C*' followed by (4) '*CpApp*' which is equiform with 8.42 after '*p*' is substituted for '*q*', and (5) the theorem '*Cpp*' which is deduced by a double application of the rule of detachment (8.33).

$8.52 \times 8.23\ q/p = 8.53$

8.53. *ANpp*

8.43 *p/Np*, *q/p* = *C*8.53 – 8.54

8.54. *ApNp*

8.54 *p/Np* × 8.23 *q/NNp* = 8.55

8.55. *CpNNp*

8.44 *p/Np*, *q/NNNp*, *r/p* = *C*8.55 *p/Np*, – *C*8.54 – 8.56

8.56. *ApNNNp*

8.43 *q/NNNp* × 8.23 *p/NNp*, *q/p* = *C*8.56 – 8.57

8.57. *CNNpp*

8.44 *q/NNp*, *r/Nq* = *C*8.55 – 8.58

8.58. *CANqpANqNNp*

8.51 *p/ANqNNp*, *q/ANNpNq*, *r/ANqp* = *C*8.43 *p/Nq*, *q/NNp* – *C*8.58 – 8.59

8.59. *CANqpANNpNq*

8.59 *p/q*, *q/p* × 8.23 × 8.23 *p/Nq*, *q/Np* = 8.60

8.60. *CCpqCNqNp*

8.41 *p/Np* × 8.23 *q/Np* = 8.61

8.61. *CCpNpNp*

8.51 *p/Apq*, *q/Aqp*, *r/p* = *C*8.43 – *C*8.42 – 8.62

8.62. *CpAqp*

8.62 *q/Nq* × 8.23 *p/q*, *q/p* = 8.63

8.63. *CpCqp*

8.63 *q/Np* = 8.64

8.64. *CpCNpp*

8.44 *p/r*, *q/Apr*, *r/q* = *C*8.62 *p/r*, *q/p* – 8.65

8.65. *CAqrAqApr*

8.44 *p/Aqr*, *q/AqApr*, *r/p* = *C*8.65 – 8.66

8.66. *CApAqrApAqApr*

8.51 *p/ApAqApr*, *q/AAqAprp*, *r/ApAqr* = *C*8.43 *q/AqApr* – *C*8.66 – 8.67

8.67. *CApAqrAAqAprp*

8.51 *p/Apr*, *q/AqApr*, *r/p* = *C*8.62 *p/Apr* – *C*8.42 *q/r* – 8.68

8.68. *CpAqApr*

8.44 *q/AqApr*, *r/AqApr* = *C*8.68 – 8.69

8.69. *CAAqAprpAAqAprAqApr*

8.51 *p/AAqAprAqApr*, *q/AqApr*, *r/AAqAprp* = *C*8.41 *p/AqApr* – *C*8.69 – 8.70

8.70. *CAAqAprpAqApr*
8.51 *p/AAqAprp, q/AqApr, r/ApAqr* = *C*8.70 – *C*8.67 – 8.71

8.71. *CApAqrAqApr*
8.44 *p/Aqr, q/Arq, r/p* = *C*8.43 *p/q, q/r* = 8.72

8.72. *CApAqrApArq*
8.51 *p/ApArq, q/ArApq, r/ApAqr* = *C*8.71 *q/r, r/q* – *C*8.72 – 8.73

8.73. *CApAqrArApq*
8.51 *p/ArApq, q/AApqr, r/ApAqr* = *C*8.43 *p/r, q/Apq* – *C*8.73 – 8.74

8.74. *CApAqrAApqr*
8.51 *p/AqApr, q/AqArp, r/ApAqr* = *C*8.72 *p/q, q/p* – *C*8.71 – 8.75

8.75. *CApAqrAqArp*
8.51 *p/ArApq, q/ArAqp, r/ApAqr* = *C*8.72 *p/r, q/p, r/q* – *C*8.73 – 8.76

8.76. *CApAqrArAqp*

HISTORY: The axiomatization of the logic of sentences was undertaken by Frege and Peano and completed in PM, which employs five axioms. This number was reduced to four by Hilbert, to three by Łukasiewicz, to one by Nicod, and this one was notably shortened by Łukasiewicz and Sobocinski.

LITERATURE: The system expounded in § 8 is that of Hilbert-Ackerman, but the method of deduction, which is not very rigorous in these authors, has been replaced by that of Łukasiewicz. The definitions are based on a discovery of Sheffer. For the systems that have been developed, see the text-books of § 0. The use of truly rigorous methods is still somewhat rare in the literature of this kind.

§ 9. A SYSTEM OF THE RULES OF DEDUCTION

The theory explained in this chapter shows how it is possible to translate logical laws into metalogical rules. In practice, the rules, which show how to proceed in deduction, are much more important than the laws, which declare not what can be done, but what is. For example, the *modus ponendo ponens* (6.42) declares that if $p \supset q$ and p, then q, but it does not at all permit one to pass from the assertion of '$p \supset q$' and of 'p' to the assertion of 'q'. However, by means of a certain few principles each law can be translated into a rule. These principles are given here without justification, although that is not hard to give.

9.1. *Definitions*

9.11. '*System 8*' for: 'system explained in § 8 and certain theorems given in §§ 5 and 6'.
9.12. '*Expression 8*' for: 'Expression of System 8'.
9.13. '*Law 8*' for: 'law of system 8'.
9.14. '*Rule* 9' for: 'rule obtained by the application of the principles of § 9 to laws 8'.

9.2. *Names of the expressions 8:*

9.21. '*Negation of X*' or '*not-X*' for: 'group composed of '*N*' and *X*'.
9.22. '*Alternation X – Y*' or '*alternation of X and Y*' for: 'group composed of '*A*', *X*, and *Y*'.
9.23. '*Implication X – Y*' or '*implication of Y by X*' for: 'group composed of '*C*', *X*, and *Y*'.
9.24. '*Disjunction X – Y*' or '*disjunction of X and Y*' for: 'group composed of '*D*', *X* and *Y*'.
9.25. '*Equivalence X – Y*' or '*equivalence of X and Y*' for: 'group composed of '*E*', *X*, and *Y*'.
9.26. '*Conjunction X–Y*' or '*conjunction of X and Y*' for: 'group composed of '*K*', *X*, and *Y*'.
Remark: The letters '*X*' and '*Y*' are variables for which only the *names* of expressions 8 can be substituted.

9.3. *Rules of Translation*

9.31. If *X* is a law 8, the expression composed of (1) the name of *X* and (2) of 'can be asserted' is a rule 9.

9.32. If the equivalence of *X* and *Y* is a law 8, the expression is a rule 9 which is composed successively of (1) the name of *X*, (2) of 'can be substituted for', (3) of the name of *Y*, (4) of 'and inversely'.
9.33. If the implication *X* – *Y* is a law 8, the expression is a rule 9 which is composed successively of (1) 'if', (2) of the name of *X*, (3) of 'is asserted, then', (4) of the name of *Y*, (5) of 'can be asserted'.
9.34. If the implication of the implication *Y* – *Z* by *X* is a law 8, then the expression is a rule 9 which is composed successively of (1) 'if', (2) of the name of *X*, (3) of 'is asserted and', (4) of the name of *Y*, (5) of 'is asserted, then', (6) of the name of *Z*, (7) of 'can be asserted'.
9.35. If the implication of *Z* by the conjunction *X* – *Y* is a law 8, then the expression is a rule 9 which is composed successively of (1) 'if', (2) of the name of *X*, (3) of 'is asserted and', (4) of the name of *Y*, (5) of 'is asserted then', (6) of the name of *Z*, (7) of 'can be asserted'.

9.4. *Examples of Rules* 9

9.41. For the negation of the negation of *X* we may substitute *X* (5.12).
9.42. For the alternation *X* – *Y* we may substitute *Y* – *X* (5.22).
9.43. For the negation of the alternation *X* – *Y* we may substitute the conjunction of *not-X* and *not-Y*. (5.27).
9.44. For the implication *X* – *Y* we may substitute the implication *not-Y* – *not-X*. (5.32).
9.45. For the implication of the implication *Y* – *Z* by *X* we may substitute the implication of the implication *X* – *Z* by *Y* (5.33).
9.46. If *X* is asserted, the alternation *X* – *Y* can be asserted (6.26).
9.47. If the implication *X* – *Y* is asserted and *X* is asserted, *Y* can be asserted (6.42).
9.48. If the implication *X* – *Y* is asserted and the negation of *Y* is asserted, the negation of *X* can be asserted (6.44).

9.5. *The schematic notation and method of Gentzen*

9.51. The rules 9 can be represented schematically by translating the expressions utilized in 9.2. and 9.3 as follows:
9.511. '$-X$' for: 'not–*X*'.
9.512. '$X + Y$' for: 'alternation *X* — *Y*'.
9.513. '$X \rightarrow Y$' for: 'implication *X* — *Y*'.
9.514. '$X = Y$' for: 'equivalence *X* — *Y*'.

9.515. 'X / Y' for: 'disjunction $X \; - Y$'.
9.516. '$X \times Y$' for: 'conjunction $X - Y$'.
9.517. '$\vdash X$' for: 'X is asserted'.
9.5171. '$\vdash X \vdash Y$' for: 'X is asserted and Y is asserted'.
9.518. '$X \infty Y$' for: 'for X we may substitute Y'.
9.519. '$\frac{\vdash X}{\vdash Y}$' for: 'if X is asserted, then Y can be asserted'.

Rules 9.42–48 can be written in this notation as follows with corresponding end ciphers 9.52–58:

9.52. $X + Y \cdot \infty \cdot Y + X$.
9.53. $- \cdot X + Y \cdot \infty \cdot - X \times - Y$.
9.54. $X \rightarrow Y \cdot \infty \cdot - Y \rightarrow - X$.
9.55. $X \cdot \rightarrow \cdot Y \rightarrow Z : \infty : Y \cdot \rightarrow \cdot X \rightarrow Z$.
9.56. $\dfrac{\vdash X}{\vdash X + Y}$
9.57. $\dfrac{\vdash \cdot X \rightarrow Y \vdash X}{\vdash Y}$
9.58. $\dfrac{\vdash \cdot X \rightarrow Y \vdash - Y}{\vdash - X}$

9.59. By establishing a small number of rules of this type (9.4 or 9.5) all of system 8 can be constructed without axioms and without recourse to the method given in § 3.

HISTORY: The distinction between laws and rules seems to go back to Husserl. The indispensability of 'rules of procedure' for the construction of a calculus was especially emphasized by Dingler. It is interesting that Aristotle considered his theorems as laws, whereas the Stoics and Scholastics took them as rules. The most notable works on rules are those of Gentzen (1934) and Jaśkowski (1934). The above elaboration of this idea is based on more recent works of the metalogicians (cf. § 26.3).

LITERATURE: Gentzen 1; Jaśkowski 1; Carnap 3; Feys 6; Popper.

III

THE LOGIC OF PREDICATES AND CLASSES

A. *The Logic of Terms*

§ 10. SYLLOGISTIC

This chapter considers the highpoint of 'classical' logic, syllogistic, which is a simple system, but one that is very important in practice. It is a system of what is called the logic of 'terms', i.e. for the variables which appear in it we can substitute terms only, and not sentences. It can be axiomatized on the basis of the sentential calculus with the help of some special axioms and 'syllogistic functors'.

10.0. *Primitive Terms and Rules*

10.001. *Primitive terms:* (a) all the primitive and defined terms of system 8 (§ 8.11–12); (b) '*a*', '*b*', and '*m*' – nominal variables, i.e. variables for which only names can be substituted; (c) dyadic functors '*A*' and '*I*' – the syllogistic functors, whose arguments are the letters '*a*', '*b*', and '*m*'. Explanation: '*a*' will be used for the major term, '*b*' for the minor term, '*m*' for the middle term. '*A*' and '*I*' (as also '*E*' and '*O*', cf. 10.01) are similar in meaning to that in classical logic, where they show the quantity and quality of a proposition. It should be noted that '*A*' and '*E*' are here *nominal* functors and hence are totally different from those defined in § 3. To avoid confusion the *sentential* functors '*A*' and '*E*' will not be used in this chapter.
10.002. *Sentences:* (1) All the sentences of system 8. (2) Groups composed of '*A*', '*E*', '*I*', or '*O*' and two of the letters '*a*', '*b*', '*m*'. (3) Sentences of system 8 in which sentences have been substituted for the variables.
10.003. *Rules:* 8.31–32–33.
10.004. *Rule:* For a variable '*a*', '*b*', or '*m*' we may substitute '*a*', '*b*', or '*m*'. This rule enables us to change the letters.
Explanation: In stating this rule as well as 8.31 it is often specified that we must not substitute for the variables the names of empty classes (cf. 15.42). With regard to this, however, it should be remarked: (1) that

this problem has nothing to do with the structure of system 10 and bears exclusively on its interpretation (cf. 7.53); in fact, rules 8.31 and 10.004 do not allow us to substitute for the variables any other expressions than the sentences given in 10.002. (2) This problem, known as the 'problem of the empty or null class' raises philosophical questions and is extremely complicated. Cf. the literature.

10.1. *Definitions and Axioms*

10.01. '*Eba*' for: '*NIba*'
10.02. '*Oba*' for: '*NAba*'

Axioms assumed from the sentential calculus:

10.03. *Cpp*	A form of the principle of identity
10.04. *CCpNqCqNp*	Simple contraposition 1°
10.05. *CCpqCNqNp*	Simple contraposition 2°
10.06. *CCpqCCqrCpr*	Principle of the (hypothetical) syllogism
10.07. *CNNpp*	Double Negation
10.08. *CCKpqrCKNrqNp*	Law for indirect reduction 1°
10.09. *CCKpqrCCspCKsqr*	Law for direct reduction 1°
10.10. *CCKpqrCpCqr*	Law of exportation 1°
10.11. *CCKpqrCqCpr*	Law of exportation 2°
10.12. *CCKpqrCCsqCKpsr*	Law for direct reduction 2°
10.13. *CCKpqrCKpNrNq*	Law for indirect reduction 2°
10.14. *CCKpqrCKqpr*	Syllogistic commutation

Remark: All these axioms are theorems demonstrable in the sentential calculus.

Special axioms:

10.15. *Aaa*	('all *a*'s are *a*'s')
10.16. *Iaa*	('some *a*'s are *a*'s')
10.17. *CKAmaAbmAba*	(Barbara)
10.18. *CKEmaIbmOba*	(Ferio)

10.2. *Logical Square and Conversion*

In the following deductions the integral part of the numeral, i.e. the '10' in '10.20', will be omitted to simplify the probative verses, and also zero when it appears in the decimal part, i.e. '10.0' in '10.03'. In the proofs the 'I°' and 'II°' indicate to which part of the expression the definition must be applied.

Laws of Contradiction

3 *p*/*Eba* × 1 II° = 20

10.20. *CEbaNIba*

3 *p*/*NIba* × 1 II° = 21

10.21. *CNIbaEba*

4 *p*/*Eba*, *q*/*Iba* = *C*20 – 22

10.22. *CIbaNEba*

5 *p*/*NIba*, *q*/*Eba* = *C*21 – (1)

(1) *CNEbaNNIba*

6 *p*/*NEba*, *q*/*NNIba*, *r*/*Iba* = *C*(1) – *C*7 *p*/*Iba* – 23

10.23. *CNEbaIba*

3 *p*/*Oba* × 2 II° = 24

10.24. *CObaNAba*

3 *p*/*Oba* × 2 I° = 25

10.25. *CNAbaOba*

4 *p*/*Oba*, *q*/*Aba* = *C*24 – 26

10.26. *CAbaNOba*

5 *p*/*NAba*, *q*/*Oba* = *C*25 – (1)

(1) *CNObaNNAba*

6 *p*/*NOba*, *q*/*NNAba*, *r*/*Aba* = *C*(1) – *C*7 *p*/*Aba* – 27

10.27. *CNObaAba*

To prove the other laws of the logical square and those of conversion it is first necessary to deduce *Datisi:*

8 *p*/*Eba*, *q*/*Imb*, *r*/*Oma* = *C*18 *b*/*m*, *m*/*b* – (1)

(1) *CKNOmaImbNEba*

9 *p*/*NOma*, *q*/*Imb*/ *r*/*NEba*, *s*/*Ama* = *C*(1) – *C*26 *b*/*m* – (2)

(2) *CKAmaImbNEba*

6 *p*/*KAmaImb*, *q*/*NEba*, *r*/*Iba* = C(2) – *C*23 – 30

10.30. *CKAmaImbIba* (Datisi)

10 *p*/*Abb*, *q*/*Iba*, *r*/*Iab* = *C*30 *a*/*b*, *b*/*a*, *m*/*b* – *C*15 *a*/*b* – 31

10.31. *CIbaIab*

11 *p*/*Aba*, *q*/*Ibb*, *r*/*Iba* = *C*30 *m*/*b* – *C*16 *a*/*b* – 32

10.32. *CAbaIba*

6 *p*/*Aba*, *q*/*Iba*, *r*/*Iab* = *C*32 – *C*31 – 33

10.33. *CAbaIab*

5 *p*/*Iab*, *q*/*Iba* × 1 × 1 *a*/*b*, *b*/*a* = *C*31 *a*/*b*, *b*/*a* – 34

10.34. *CEbaEab*

5 *p/Aba, q/Iba* × 1 × 2 = *C*32 – 35

10.35. *CEbaOba*

6 *p/Eba, q/Eab, r/Oab* = *C*34 – *C*35 *a/b, b/a* – 36

10.36. *CEbaOab*

5 *p/Aba, q/Iba* = *C*32 – 37

10.37. *CNIbaNAba*

5 *p/Eba, q/Oba* = *C*35 – 38

10.38. *CNObaNEba*

6 *p/Aba, q/NOba, r/NEba* = *C*26 – *C*38 – 39

10.39. *CAbaNEba*

6 *p/Eba, q/NIba, r/NAba* = *C*20 – *C*37 – 40

10.40. *CEbaNAba*

6 *p/NIba, q/NAba, r/Oba* = *C*37 – *C*25 – 41

10.41. *CNIbaOba*

6 *p/NOba, q/NEba, r/Iba* = *C*38 – *C*23 – 42

10.42. *CNObaIba*

In addition to the laws of conversion (10.31–33–34–36), there are others for obversion, contraposition, etc., which are frequently studied. They can be deduced in the system by adding two axioms and certain definitions, but since their practical and theoretical importance is slight, they have been omitted.

10.5. *The Moods of the Syllogism*

6 *p/KAmaAbm, q/Aba, r/Iba* = *C*17 – *C*32 – 50

10.50. *CKAmaAbmIba* (Barbari)

12 *p/Ama, q/Imb/ r/Iba, s/Ibm* = *C*30 – *C*31 *a/m* – 51

10.51. *CKAmaIbmIba* (Darii)

9 *p/Ema, q/Ibm, r/Oba, s/Eam* = *C*18 – *C*34 *a/m, b/a* – 52

10.52. *CKEamIbmOba* (Festino)

13 *p/Ema, q/Iba, r/Obm* × 1 = *C*52 *a/m, m/a* – (1)

(1) *CKEmaNObmEba*

12 *p/Ema, q/NObm,* r*/Eba, s/Abm* = *C*(1) – *C*26 *a/m* – 53

10.53. *CKEmaAbmEba* (Celarent)

6 *p/KEmaAbm, q/Eba, r/Oba* = *C*53 – *C*35 – 54

10.54. *CKEmaAbmOba* (Celaront)

13 *p/Aam, q/Aba, r/Abm* × 2 *a/m* × 2 = *C*17 *a/m, m/a* – 55

10.55. *CKAamObmOba* (Baroco)
9 *p/Ema, q/Abm, r/Eba, s/Eam* = *C*53 – *C*34 *a/m, b/a* – 56

10.56. *CKEamAbmEba* (Cesare)
6 *p/KEamAbm, q/Eba, r/Oba* = *C*56 – *C*35 – 57

10.57. *CKEamAbmOba* (Cesaro)
14 *p/Ema, q/Abm, r/Eba* = *C*53 – (1)

(1) *CKAbmEmaEba*
12 *p/Aam, q/Emb, r/Eab, s/Ebm* = *C*(1) *a/b, b/a* – *C*34 *a/m* – (2)

(2) *CKAamEbmEab*
6 *p/KAamEbm, q/Eab, r/Eba* = *C*(2) – *C*34 *a/b, b/a* – 58

10.58. *CKAamEbmEba* (Camestres)
6 *p/KAamEbm, q/Eba, r/Oba* = *C*58 – *C*35 – 59

10.59. *CKAamEbmOba* (Camestrop)
8 *p/Aba, q/Amb, r/Ama* × 2 *b/m* × 2 = *C*17 *b/m, m/b* – 60

10.60. *CKOmaAmbOba* (Bocardo)
14 *p/Amb, q/Ima, r/Iab* = *C*30 *a/b, b/a* – (1)

(1) *CKImaAmbIab*
6 *p/KImaAmb, q/Iab, r/Iba* = *C*(1) – *C*31 *a/b, b/a* – 61

10.61. *CKImaAmbIba* (Disamis)
12 *p/Ama, q/Imb, r/Iba, s/Amb* = *C*30 – *C*32 *a/b, b/m* – 62

10.62. *CKAmaAmbIba* (Darapti)
12 *p/Ema, q/Ibm, r/Oba, s/Imb* = *C*18 – *C*31 *a/b, b/m* – 63

10.63. *CKEmaImbOba* (Ferison)
12 *p/Ema, q/Ibm, r/Oba, s/Amb* = *C*18 – *C*33 *a/b, b/m* – 64

10.64. *CKEmaAmbOba* (Felapton)
12 *p/Eam, q/Ibm, r/Oba, s/Imb* = *C*52 – *C*31 *a/b, b/m* – 65

10.65. *CKEamImbOba* (Fresison)
12 *p/Eam, q/Ibm, r/Oba, s/Amb* = *C*52 – *C*33 *a/b, b/m* – 66

10.66. *CKEamAmbOba* (Fesapo)
9 *p/Ima, q/Amb, r/Iba, s/Iam* = *C*61 – *C*31 *a/m, b/a* – 67

10.67. *CKIamAmbIba* (Dimaris)
9 *p/Ima, q/Amb, r/Iba, s/Aam* = *C*61 – *C*33 *a/m, b/a* – 68

10.68. *CKAamAmbIba* (Bamalip)
12 *p/Aam, q/Ebm, r/Eba, s/Emb* = *C*58 – *C*34 *a/b, b/m* – 69

10.69. *CKAamEmbEba* (Camenes)
6 *p/KAamEmb, q/Eba, r/Oba* = *C*69 – *C*35 – 70

10.70. *CKAamEmbOba* (Camenop)

HISTORY: Syllogistic is a discovery of Aristotle. It was further developed by his followers and the Scholastics, to whom we owe the mnemonic verse, 'Barbara, Celarent, etc.' A rigorous axiomatization of syllogistic was first undertaken by Łukasiewicz in 1929.

LITERATURE: The best non-mathematical exposition is that of Keynes. History: Bocheński 7, 8. Axiomatization: Łukasiewicz 3, 7; Bocheński 3, 5; Thomas 2, 3, 4; Wedberg; Menne 4. Other methods: Ajdukiewicz 1; Black 2; Curry 3; Feys 5; Greenwood; Miller; Moisil 2.

B. *The Logic of Predicates*

§ 11. MONADIC PREDICATES

Whereas the syllogistic (cf § 10) analyses the sentence into subject and predicate and conceives both as arguments of a dyadic functor '*A*', '*I*', '*E*', or '*O*', the logic of monadic predicates conceives the predicate as a functor and the subject as its argument. The quantity of the expression is indicated by a special expression, called the 'quantifier'. Ontologically speaking, we may say that we are dealing with 'individuals', 'properties', and the extent to which the 'properties' apply to 'individuals', i.e. their 'quantity'. But logically speaking, we consider the predicate only as a nominal functor which with a name as argument forms a sentence whose quantity is indicated by the 'quantifier'.

11.1. *Definitions*

11.11. '*Individual constant*' for: 'letter 'a', 'b', 'c', or 'd'.'
11.12. '*Individual variable*' for: 'letter 'x', 'y', 'z', or 't'.'
11.13. '*Individual functor*' for: 'letter 'φ', 'ψ', 'χ', 'θ'.'
11.14. '*Individual sentence*' for: 'expression composed of an individual functor and individual constants'.
Explanation: 'φa' is an individual sentence, which is read as 'φ of a' and which signifies that the property φ belongs to the individual a.
11.15. '*Matrix*' for: 'individual functor followed by individual variables'.
Explanation: 'φx' is a matrix. It is not a sentence, but it can become one if an individual constant is substituted for the variable or if the expression is quantified.

11.2. *Quantifiers*

11.21. '*Universal quantifier*' for: 'one or more variables, separated by commas, between round parentheses, in the Peano-Russell notation, or preceeded by 'Π' in the Łukasiewicz notation, the whole placed before a matrix.'
Explanation: In '$(x)\varphi x$' or '$\Pi x\varphi x$' the '(x)' or 'Πx' is the universal quantifier. The whole is read: 'for all x: φ of x'; for example if 'φ' is 'smokes', we have 'for all x: smokes of x', i.e. 'everything smokes'. Note that when the matrix is so quantified, it becomes a sentence, since it is true or false.

11.22. '*Existential quantifier*' for: ''E' followed by one or more variables separated by commas, between round parentheses, in the Peano-Russell notation, or proceeded by 'Σ' in the Łukasiewicz notation, the whole placed before a matrix'.
Explanation: In '$(Ex)\varphi x$' or '$\Sigma x \varphi x$' the '(Ex)' or 'Σx' is the existential quantifier. The whole is read: 'there is at least one x such that φ of x'; for example, if 'φ' is 'smokes', we have 'there is at least one x such that smokes of x', i.e. 'there is one being that smokes'.
N.B. In PM the 'E' of the existential quantifier is reversed, thus '$(\exists)$'.
11.23. 'Quantifier' for: 'universal quantifier or existential quantifier'.

11.3. *Free and Bound Variables*

11.31. '*Free variable*' or '*real variable*' for: 'variable contained in a matrix not preceeded by a quantifier which contains a letter of the same shape'.
Example: the variable 'x' in '$\varphi x \supset \psi x$'.
11.32. '*Bound variable*' or '*apparent variable*' for: 'variable contained in a matrix which is preceeded by a quantifier containing a letter of the same shape as the variable'.
Example: 'x' is a bound variable in: '$(x)\varphi x \supset \psi x$', since the matrix in question is preceeded by '(x)'.
11.33. *Rule:* Substitution cannot be made for a bound variable.
11.34. 'X is bound by the quantifier Y' for: 'X is a variable which is part of a matrix preceeded by Y, and Y contains a letter of the same shape as X'.
11.35. *Rule:* No variable can be bound by more than one quantifier.
11.36. '*Universal closure of* X' or '*universalization of* X' for: 'expression of the same shape as X, preceeded by universal quantifiers binding all the variables of X, where X is a matrix'.
11.37. '*Existential closure of* X' or '*particularization of* X' for: 'expression of the same shape as X, preceeded by existential quantifiers binding all the variables of X, where X is a matrix'.
11.38. '*Closure of* X' or '*generalization of* X' for: 'universalization or particularization of X'.
Examples: '$(x)\varphi x$' is a universal closure or universalization of 'φx'. '$(Exy) \cdot \varphi x \cdot \psi y$' an existential closure or particularization of $\varphi x \cdot \psi y$. Both are generalizations.

Explanation: A closure is not a matrix, but a sentence; there can be no substitution for its variables; it has a value, whereas the matrix does not. – The laws of 8 should all be preceeded by quantifiers; if they are omitted it is because in the logic of sentences all the quantifiers are universal, and there is no risk of equivocation. Nevertheless, it is possible even in this domain to construct a theory with existential quantifiers.

11.39. '*Formal implication*' for: 'universal closure of an expression composed of a matrix, of '$\supset$', and of another matrix, where the variables of the first matrix are of the same shape as those of the second'.

Example: '$(x) \cdot \varphi x \supset \psi x$'.

Explanation: Formal implication (with constant functors) corresponds more or less to the universal affirmative sentence of ordinary language: 'All logicians are pipe-smokers' can be written: '$(x) \cdot$ logician $(x) \supset$ pipe-smoker (x)'. – Thus there are three implications to be distinguished: material implication (3.5), formal implication (11.39), and inferential implication (7.48).

HISTORY: The analysis of a sentence into a predicate functor and its argument as well as formal implication is found in Aristotle; by the time of Albert the Great it was further developed and used in modal logic. However, the idea of writing the sentence as a function with constant use of quantifiers and the invention of a notation for it is the work of Frege 1. This was the decisive step towards the formation of formal logic in its present state. The theory has recently received considerable development in the 'combinatory' logic founded by Schönfinkel and Curry (cf. § 26.2).

LITERATURE: The classical theory of the predicate calculus and the enumeration of its laws is found in all good textbooks, in particular in PM *9 – *10.

§ 12. LAWS OF MONADIC PREDICATES

This chapter contains, without demonstration, the most fundamental laws of the logic of monadic predicates. These laws form the basis of the logical theories that follow.
In this and the following chapters we will sometimes use a greater number of points than is strictly necessary in order to facilitate understanding.

12.1. *Methodological principle*

All the laws of monadic predicates can be deduced from the laws 8 and the two following definitions:
12.11. '$(x)\varphi x$' for: '$\varphi a \cdot \varphi b \cdot \varphi c \cdot \varphi d \cdot \ldots$',
12.12. '$(Ex)\varphi x$' for: '$\varphi a \vee \varphi b \vee \varphi c \vee \varphi d \vee \ldots$', where the number of arguments is taken as indefinite.
Explanation: 12.11 supposes that 'all x's possess the property φ' signifies the same thing as 'a possesses the property φ, and b possesses it, and c, etc.'. 12.12 says that 'some x possesses the property φ' signifies that 'a possesses the property φ, or b possesses it, or c, etc.'. These definitions run into very serious logical difficulties, since the notion of 'etc' is very complicated and cannot be defined without the aid of expressions of the type used here. But they are useful in practice. Furthermore, the great majority of the laws of predicates can be deduced by still more restricted definitions:

'$(x)\varphi x$' for: '$\varphi a \cdot \varphi b$'
'$(Ex)\varphi x$' for: '$\varphi a \vee \varphi b$'.

In fact, all the sentences deduced from these definitions by the use of rules 9 are true, as long as individual constants are not introduced.

12.2. *Negation of quantified monadic predicates*

12.21. $(x)\varphi x \equiv \sim (Ex) \sim \varphi x$ $\quad E\Pi x\varphi xN\Sigma xN\varphi x$
12.22. $\sim (x)\varphi x \equiv (Ex) \sim \varphi x$ $\quad EN\Pi x\varphi x\Sigma xN\varphi x$
12.23. $(x) \sim \varphi x \equiv \sim (Ex)\varphi x$ $\quad E\Pi xN\varphi xN\Sigma x\varphi x$
12.24. $\sim (x) \sim \varphi x \equiv (Ex)\varphi x$ $\quad EN\Pi xN\varphi x\Sigma x\varphi x$

12.25. *Rule:* Negating all the quantifiers and matrices and substituting existential for universal quantifiers, and inversely, does not change the value of the sentence.

12.3. *Fundamental laws*

12.31. $(x)\varphi x \supset \varphi y$ $\qquad$ $C\Pi x\varphi x\varphi y$
12.32. $\varphi y \supset (Ex)\varphi x$ $\qquad$ $C\varphi y\Sigma x\varphi x$
12.33. $(x)\varphi x \supset (Ex)\varphi x$ $\qquad$ $C\Pi x\varphi x\Sigma x\varphi x$
Explanation: 12.31 signifies: 'if φ (universally) of all x, then φ of y', this law is deduced from 12.11 by 6.27. 12.32 signifies: 'if φ of y, then there is at least one x such that φ of x'; it is deduced from 12.12 by 6.26. 12.33 is the well known law of subalternation, which is obtained by the law of the syllogism (6.31 sq.) from 12.31 and 12.32.
12.34. Laws 12.31–32 added as axioms to system 8, with certain new rules and definitions, suffice to establish the axiomatic system of predicates.

12.4. *Rules of deduction*

12.41. The universal quantifier placed at the beginning of an asserted sentence can be omitted if it extends to all the expressions which follow in the sentence.
12.411. If the universal closure of the matrix X is asserted, the expression formed by substituting constants for the variables of X can be asserted (12.31).
Example: Take as asserted 'all x's are moral', i.e. '(x) mortal x'. Then by 12.31 the sentence 'Peter is mortal' can be asserted.
12.42. If the matrix X is asserted, the existential closure of X can be asserted (12.32).
12.421. If the individual sentence X is asserted, the existential closure of the matrix formed by substituting variables for the constants of X can be asserted (12.32).
Example: Take the individual sentence 'Peter smokes', i.e. 'smokes (Peter)'. Then by 12.32 '(Ex) smokes x', i.e. the sentence 'there is one x which smokes' can be asserted.
12.43. If the universal closure of X is asserted, the existential closure of X can be asserted (12.33).
12.44. If the matrix X is asserted, the universal closure of X can be asserted.
Explanation: 12.44 is not founded on a law, as are 12.41–42–43. But it can be justified either by the application of method 12.1 or by the

following consideration: 'φx' asserts that φx belongs to any x; φ then belongs to all x; which is what is expressed by '$(x)\varphi x$'.

12.5. *Analogous laws*

12.51. 'X is an expression analogous (12.5) to Y' for: 'X is an expression formed from Y by substituting 'φx' for 'p', 'ψx' for 'q', 'χx' for 'r', 'θx' for 's' and preceeded by '(x)' or by 'Πx''.
12.52. Every expression analogous to a law 8 is a law.
12.53. $(x) \cdot \varphi x \equiv \varphi x$ — $\Pi x E \varphi x \varphi x$
Principle of identity for predicates (cf. 5.11)
12.54. $(x) \cdot \varphi x \mid \sim \varphi x$ — $\Pi x D \varphi x N \varphi x$
12.55. $(x) : \sim \cdot \varphi x \cdot \sim \varphi x$ — $\Pi x N K \varphi x N \varphi x$
Principle of non-contradiction for predicates (cf. 6.11–12)
12.56. $(x) \cdot \varphi x \vee \sim \varphi x$ — $\Pi x A \varphi x N \varphi x$
Principle of excluded middle for predicates (cf. 6.13)
12.57. $(x) :\cdot \varphi x \supset \psi x : \supset : \psi x \supset \chi x \cdot \supset \cdot \varphi x \supset \chi x$
$\Pi x C C \varphi x \psi x C C \psi x \chi x C \varphi x \chi x$ (cf. 6.32)
Principle of syllogism for predicates
12.58. $(x) : \varphi x \supset \psi x \cdot \varphi x \cdot \supset \cdot \psi x$ — $\Pi x C K C \varphi x \psi x \varphi x \psi x$
Modus ponendo ponens for predicates (cf. 6.42).

12.6. *Laws for the movement of quantifiers*

12.61. $(x) \cdot \varphi x \cdot \psi x \cdot \equiv \cdot (\chi)\varphi x \cdot (x)\psi x$ — $E \Pi x K \varphi x \psi x K \Pi x \varphi x \Pi x \psi x$
Example: If all men are mammals and bipeds, then all men are mammals and all men are bipeds. The inverse is true.
12.62. $(Ex) \cdot \varphi x \cdot \psi x \cdot \supset \cdot (Ex)\varphi x \cdot (Ex)\psi x$ — $E \Sigma x K \varphi x \psi x K \Sigma x \varphi x \Sigma x \psi x$
Example: If there is a man who is a logician and a pipe smoker, there is a man who is a logician and there is a man who is a pipe-smoker; the inverse is not true.
12.63. $(Ex) \cdot \varphi x \vee \psi x \cdot \equiv \cdot (Ex)\varphi x \vee (Ex)\psi x$ — $E \Sigma x A \varphi x \psi x A \Sigma x \varphi x \Sigma x \psi x$
12.64. $(x)\varphi x \cdot \vee \cdot (x)\psi x : \supset : (x) \cdot \varphi x \vee \psi x$ — $C A \Pi x \varphi x \Pi x \psi x \Pi x A \varphi x \psi x$

Example: If all locomotives are large or all locomotives are small, all locomotives are large or small. The inverse is not true.

12.65. $(x) \cdot \varphi x \supset \psi x : \supset : (x)\varphi x \cdot \supset \cdot (x)\psi x$ $\quad C\Pi xC\varphi x\psi xC\Pi x\varphi x\Pi x\psi x$

12.66. $(x) \cdot \varphi x \equiv \psi x : \supset : (x)\varphi x \cdot \equiv \cdot (x)\psi x$ $\quad C\Pi xE\varphi x\psi xE\Pi x\varphi x\Pi x\psi x$

The inverse is not true.

The following laws, under 12.7, allow for the movement of the quantifier when there is a sentence '*p*' not containing '*x*'.

12.71. $(x) \cdot \varphi x \vee p : \equiv : (x)\varphi x \cdot \vee \cdot p$ $\quad E\Pi xA\varphi xpA\Pi x\varphi xp$

12.72. $(Ex) \cdot \varphi x \vee p : \equiv : (Ex)\varphi x \cdot \vee \cdot p$ $\quad E\Sigma xA\varphi xpA\Sigma x\varphi xp$

12.73. $(x) \cdot p \supset \varphi x : \equiv : p \cdot \supset \cdot (x)\varphi x$ $\quad E\Pi xCp\varphi xCp\Pi x\varphi x$

12.74. $(Ex) \cdot p \supset \varphi x : \equiv : p \cdot \supset \cdot (Ex)\varphi x$ $\quad E\Sigma xCp\varphi xCp\Sigma x\varphi x$

On the other hand we also have:

12.75. $(x) \cdot \varphi x \supset p : \equiv : (Ex)\varphi x \cdot \supset \cdot p$ $\quad E\Pi xC\varphi xpC\Sigma x\varphi xp$

12.76. $(Ex) \cdot \varphi x \supset p : \equiv : (x)\varphi x \cdot \supset \cdot p$ $\quad E\Sigma xC\varphi xpC\Pi x\varphi xp$

Explanation: The apparent paradox of these last laws, by which we have the equivalence of universal and existential sentences, disappears upon consideration of 12.21, 12.22, and 5.311.

12.8. *Syllogistic laws*

12.81. $(x) \cdot \varphi x \supset \psi x : (x) \cdot \chi x \supset \varphi x : \supset : (x) \cdot \chi x \supset \psi x$
$CK\Pi xC\varphi x\psi x\Pi xC\chi x\varphi x\Pi xC\chi x\psi x$ (cf. 10.17)

12.82. $(x) \cdot \varphi x \supset \psi x : (Ex)\varphi x : \supset : (Ex)\psi x$
$CK\Pi xC\varphi x\psi x\Sigma x\varphi x\Sigma x\psi x$ (cf. 10.51)

12.83. $(x) \cdot \varphi x \supset \psi x : (Ex) \sim \psi x : \supset : (Ex) \sim \varphi x$
$CK\Pi xC\varphi x\psi x\Sigma xN\psi x\Sigma xN\varphi x$ (cf. 10.55)

12.84. $(x) \cdot \varphi x \vee \psi x : (Ex) \sim \varphi x : \supset : (Ex)\psi x$
$CK\Pi xA\varphi x\psi x\Sigma xN\varphi x\Sigma x\psi x$ (cf. 6.52)

12.85. $(x) \cdot \varphi x \mid \psi x : (Ex)\varphi x : \supset \cdot (Ex) \sim \psi x$
$CK\Pi xD\varphi x\psi x\Sigma x\varphi x\Sigma xN\psi x$ (cf. 6.54)

12.86. $(x) \cdot \varphi x \equiv \psi x : (Ex)\varphi x : \supset : (Ex)\psi x$
$CK\Pi xE\varphi x\psi x\Sigma x\varphi x\Sigma x\psi x$

12.87. $(x) \cdot \varphi x \equiv \sim \psi x : (Ex)\varphi x : \supset : (Ex) \sim \psi x$
$CK\Pi xE\varphi xN\psi x\Sigma x\varphi x\Sigma xN\psi x$

12.9. *Laws with individual constants*

12.91. $(x) \cdot \varphi x \supset \psi x : \varphi a : \supset : \psi a$ $\quad CK\Pi xC\varphi x\psi x\varphi a\psi a$

Explanation: Both 12.81 and 12.91 were represented in the traditional logic by *Barbara* (10.17), although there is a considerable difference between them.

12.92. $(x) \cdot \varphi x \supset \psi x : \sim \psi a : \supset : \sim \varphi a$ $CK\Pi xC\varphi x\psi xN\psi aN\varphi a$

Explanation: 12.92 is another form of *Baroco* (10.55); cf. 12.83.

12.93. $(x) \cdot \varphi x \vee \psi x : \sim \varphi a : \supset : \psi a$ $CK\Pi xA\varphi x\psi xN\varphi a\psi a$ (cf. 12.84)

12.94. $(x) \cdot \varphi x \mid \psi x : \varphi a : \supset : \sim \psi a$ $CK\Pi xD\varphi x\psi x\varphi aN\psi a$ (cf. 12.85)

12.95. $(x) \cdot \varphi x \equiv \psi x : \varphi a : \supset : \psi a$ $CK\Pi xE\varphi x\psi x\varphi a\psi a$ (cf. 12.86)

12.96. $(x) \cdot \varphi x \equiv \sim \psi x : \varphi a : \supset : \sim \psi a$ $CK\Pi xE\varphi xN\psi x\varphi aN\psi a$ (cf. 12.87)

12.97. The theory explained in this chapter is called the 'predicate calculus of the first order' or 'lower calculus'. There is also a 'higher calculus' which considers the predicates of predicates where the predicates themselves are quantified. This calculus, although indispensable for analysis, has not yet been developed formally.

LITERATURE: PM, Scholz 5; Hilbert A, Hilbert B; on 12.97: Hilbert A; Chwistek 3; Ackermann 1; Bernays 1; Quine 5.

§ 13. DYADIC PREDICATES

In the sciences as in daily life we often employ dyadic predicates (for example, 'Isodore smokes a pipe') and, what is more important, with both arguments quantified, as for example in the sentence, 'there are men who love all living things'. The theory of these predicates is readily obtained from the basis afforded by § 12.

13.1. *Definitions*

13.11. '$\varphi(x, y)$' for: 'φ of x and y'.
13.12. '$(x, y)\varphi x(x, y)$' for: '$(x) \cdot (y)\varphi(x, y)$'.
'$\Pi xy\varphi yx$' for: '$\Pi x\Pi y\varphi xy$'.
13.13. '$(Ex, y)\varphi(x, y)$' for: '$(Ex) \cdot (Ey)\varphi(x, y)$'.
'$\Sigma xy\varphi xy$' for: '$\Sigma x\Sigma y\varphi xy$'.
13.14. '$(x)(Ey)\varphi(x, y)$' for: '$(x) \cdot (Ey)\varphi(x, y)$'.
13.15. '$(Ex)(y)\varphi(x, y)$' for: '$(Ex) \cdot (y)\varphi(x, y)$'.

13.2. *Laws for the movement of quantifiers*

13.21. $(x, y)\varphi(x, y) \cdot \equiv \cdot (y, x)\varphi(x, y)$ $\quad E\Pi xy\varphi xy\Pi yx\varphi xy$
13.22. $(Ex, y)\varphi(x, y) \cdot \equiv \cdot (Ey, x)\varphi(x, y)$ $\quad E\Sigma xy\varphi xy\Sigma yx\varphi xy$
13.23. *Rule:* If the quantifiers of a sentence binding the arguments of the same individual functor are all universal or all existential, their order can be changed without changing the value of the sentence.
13.24. $(Ex)(y)\varphi(x, y) \cdot \supset \cdot (y)(Ex)\varphi(x, y)$ $\quad C\Sigma x\Pi y\varphi xy\Pi y\Sigma x\varphi xy$
Explanation: This law is only an implication, and not an equivalence, since its inverse:

$$(x)(Ey)\varphi(x, y) \cdot \supset \cdot (Ey)(x)\varphi(x, y)$$

is false, as can be seen from the following example. Let '$\varphi(x, y)$' be an abbreviation for 'x resembles y'. Then '$(x)(Ey)\varphi(x, y)$' reads: 'for all x there is at least one y such that x resembles y', i.e. 'each thing has something which resembles it'. But '$(Ey)(x)\varphi(x, y)$' reads: 'there exists at least one y such that, for all x, x resembles y', i.e. 'there is at least one thing which resembles everything'. The first sentence seems to be true, while the second is manifestly false.

13.3. *Analogous laws*

13.31. 'X is an expression analogous (13.3) to Y' for: 'X is an expression formed by substituting for all 'x' in Y '(x, y)' in the arguments and 'x, y' in the quantifiers'.

13.32. Every expression analogous (13.3) to a law of § 12 is a law.

13.33. *Rule:* By constructing a definition like 13.31 for triadic and higher functors, a rule of the same kind can be formed for the establishing analogous laws for these predicates.

HISTORY: The first appearance of the logic of dyadic predicates seems to be in the work of Frege and Peano. It is one of the most important acquisitions of mathematical logic.

LITERATURE: Hilbert A; PM *11; and the other textbooks.

§ 14. IDENTITY AND DESCRIPTION

Two rather different theories are brought together in this chapter. That of identity serves as a preamble to the logic of classes and plays a considerable role in the further developments of logic; it considers the notion 'x is the same as y'. – The theory of description is a kind of logical grammar of the definite article, 'the'. It enables us to formulate and axiomatize such expressions as 'the x such that...'. It is of great importance in the application of logic.

14.1. *Identity*

14.11. '$x = y$' for: 'x is identical with y'.
Explanation: Identity can be defined as follows:

'$x = y$' for: '$(\varphi) \cdot \varphi x \supset \varphi y$

But this definition, based on the *principle of indiscernibles* of Leibniz and called the 'thesis of extensionality', involves serious difficulties in applying logic to other domains. For this reason it is better to introduce identity as a primitive or undefined term.

14.12. '$x \neq y$' for: '$\sim \cdot x = y$'.
Explanation: 14.12. defines diversity.

14.13. $(x) \cdot x = x$.
Explanation: 14.13 is another form of the principle of identity (cf. 5.11 and 12.53).

14.14. $(x, y) : x = y \cdot \equiv \cdot y = x$.

14.15. $(x, y, z) : x = y \cdot y = z \cdot \supset \cdot x = z$.
Explanation: These three laws formulate the principal characteristics of identity: it is reflexive (14.13), symmetrical (14.14) and transitive (14.15) Cf. § 22.

14.16. 'xIy' for: '$x = y$'.

14.17. 'xJy' for: '$x \neq y$'.

14.18. $(x, y) : x = y \cdot \supset \cdot (\varphi) \cdot \varphi x \supset \varphi y$.
Explanation: If x and y are identical, y possesses all the predicates that x does.

14.2. *Descriptions*

14.21. '*Description*' for: 'a monadic matrix, preceeded by 'ι' (inverted iota) and a variable of the same shape as that in the matrix between parentheses'.

14.22. '$(\imath x)(\varphi x)$' for: 'the x such that φx'. *Description.*
Explanation: The description functor '$(\imath x)$' is like the quantifier in taking only a matrix for argument. With it an individual name is formed. Examples: if 'φ' is 'author of *Quo Vadis*', then '$(\imath x)(\varphi x)$' is 'the author of *Quo Vadis*'. In the same way we could have 'the square of 9', 'the first king of Hungary', 'John's automobile'.
14.23. '$E!(\imath x)(\varphi x)$' for: '$(Eb)(x) : \varphi x \cdot \equiv \cdot x = b : \varphi b$'.
Explanation: According to this definition, '$E!(\imath x)(\varphi x)$' signifies that the thing described by '$(\imath x)(\varphi x)$' exists and is unique; it exists, as is shown by '(Eb)'; it is unique, since according to the definition every x which possesses the property φ is identical with this b. To describe by 'the' a class which has more than one element is without meaning; for example, the expression 'the English general' without further qualification is without meaning since there is more than one English general.
14.24. $\psi[(\imath x)(\varphi x)] \cdot \supset \cdot E!(\imath x)(\varphi x)$.
Explanation: the assertion that the thing described possesses a property implies its existence. Example: 'the author of the *Divine Comedy* was Italian' implies that its author existed; 'John's automobile is a Vauxhall' implies that there is an automobile which John has.

HISTORY: The theory of identity was investigated by Leibniz and developed by Peano. The theory of description, known to Frege and Peano, was elaborated most by Russell. It involves difficult philosophical problems which have not yet been completely cleared up.

LITERATURE: § 14.1: PM *13; Scholz 5, 3; on the difficulties of Leibniz's definition PM I, p. 659 sq.; Ajdukiewicz 3 § 14.2: PM *14; Russell 2; Moore; a different point of view in Hilbert B; Quine 3.

C. *The Logic of Classes*

§ 15. CLASSES

Whereas the calculus of predicates considers the comprehension of terms (functors), the calculus of classes examines their extension. The two are perfectly analogous. The Peano-Russell system will be followed here. However, it should be noted that there is a newer theory elaborated by Leśniewski (which he calls 'ontology') which does not admit the null class and which bases itself on only one primitive term, 'is'.

15.1. *Fundamental definitions*

15.11. '$\hat{x}(\varphi x)$' for 'the x's such that: φx'.
Examples: 'The x's such that: x smokes a pipe', i.e. 'pipesmokers'; 'the x's such that: x lives in London', i.e. 'the inhabitants of London'.
Explanation: 15.11 defines a class by a sentential function; The functor '$\hat{}$', called 'Abstractor' or 'Comprehensor', has as argument a sentence, from which it forms a class. This operation is called 'abstraction': the class of pipe-smokers is an abstraction from the function 'x smokes a pipe'.
15.111. '$\lambda x \varphi x$' for: '$\hat{x}(\varphi x)$'.
Explanation: 'The expression 'λx', which is called the 'lambda operator' is often substituted, especially in recent times', for the x with circumflex used in PM.
15.12. 'Cls' for: '$\hat{\alpha}\{(E\varphi) \cdot \alpha = \hat{x}(\varphi x)$'.
Explanation: This is the definition of the class of classes: it is composed of all the α such that $\alpha = \hat{x}(\varphi x)$ for any φ, i.e. according to 15.11 for all classes.
15.13. '$y \varepsilon \hat{x}(\varphi x)$' for: '$\varphi y$'.
Explanation: To say: 'y is an element of the class of those x's for which φx holds' amounts to saying 'φy'. The 'ε' here is a dyadic functor which, in the Peano-Russell notation, is written between the arguments, and which forms a sentence. The first argument must be the name of an individual (constant or variable) and the second a class.
Example: If 'y is an element of the class of those x's for which being-a-Swiss holds for x', then we can say 'y is a Swiss'. Thus each man who smokes a pipe is an element of the class of smokers and each mountain of the Alps is an element of the class called 'Alps', etc.

15.14. '$x \sim \varepsilon a$' for: '$\sim \cdot x\varepsilon a$'.
15.15. '$x, y\varepsilon a$' for: '$x\varepsilon a \cdot y\varepsilon a$'.

15.2. *Relations between classes*

15.21. '$-a$' for: '$\hat{x}(x \sim \varepsilon a)$'. The *complementary class* of a.
Explanation: The complementary class of a includes as elements all things that are not elements of a. Example: The complementary class of the class of elephants is the class of non-elephants. It is evident that the world is full of non-elephants.
15.22. '$a \cup \beta$' for: '$\hat{x}(x\varepsilon a \cdot \vee \cdot x\varepsilon\beta)$'. The *logical sum* of classes.
15.23. '$a \cap \beta$' for: '$\hat{x}(x\varepsilon a \cdot x\varepsilon\beta)$'. The *product* of classes.
15.24. '$a \parallel \beta$' for: '$\hat{x}(x\varepsilon a \cdot \mid \cdot x\varepsilon\beta)$'. The *disjunction* of classes.
Explanation: Let a be the class of pipe-smokers and β that of logicians. In this case $a \cup \beta$ is the class of all those who are either pipe-smokers or logicians. $a \cap \beta$ is the class of logicians who are pipe-smokers. $a \parallel \beta$ is the class of those who are not both logicians and pipe-smokers.
15.25. '$a \subset \beta$' for: '$(x) : x\varepsilon a \cdot \supset \cdot x\varepsilon\beta$'. *Inclusion* of classes.
15.26. '$a = \beta$' for: '$(x) : x\varepsilon a \cdot \equiv \cdot x\varepsilon\beta$'. *Equality* of classes.
Examples: The class of pipe-smokers is included in the class of smokers; that of French citizens who are 21 or older is equal to the class of men who have the right to vote in France. Note that '$a \subset \beta$' and '$a = \beta$' are sentences whereas '$a \cup \beta$' and '$a \cap \beta$' are names of classes.
15.27. *Rule of points:* A group of points placed next to a truth functor has a higher rank than a group of points placed next to one of the functors defined in 15.21–26.

15.3. *Graphical representation*

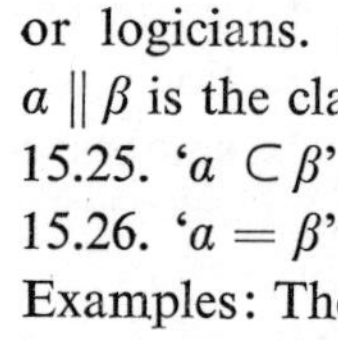

$-\ a$

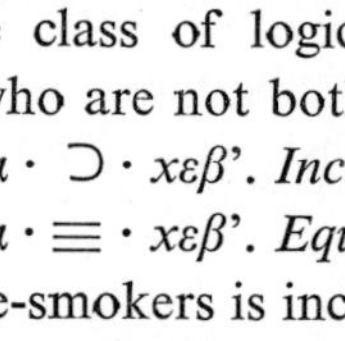

$a \cup \beta$

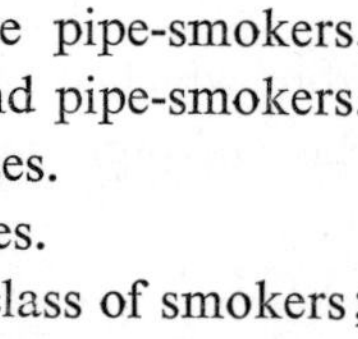

$a \cap \beta$

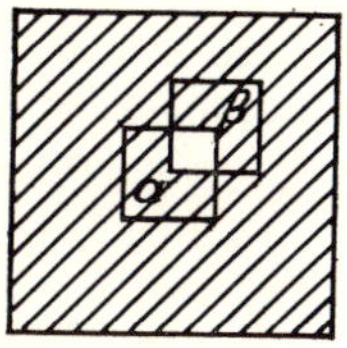

$\alpha \mid\mid \beta$

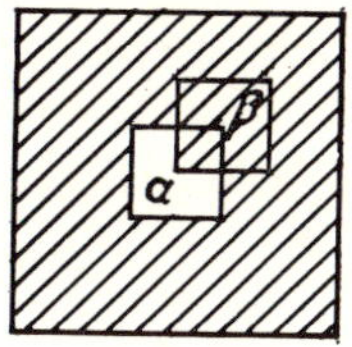

$\alpha \subset \beta$

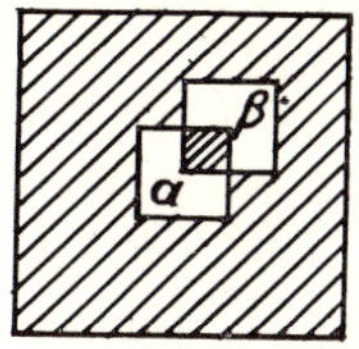

$\alpha = \beta$

15.4. *Existence*

15.41. '$\mathbf{V}$' for: '$\hat{x}(x = x)$'. *Universal class.*
15.42. 'Λ' for: '$\hat{x}(x \neq x)$'. *Null Class.*
Explanation: The universal class is the class of all x's that are identical with themselves, i.e. of all x in general, since everything is identical with itself. The null class is the class of all x's which are not identical with themselves, i.e. of objects which do not exist. Examples: the class of Swiss kings, of wives of Copernicus, of fathers of Adam, of the automobiles of a man who has none, all belong to the null class.
15.43. '$\exists!a$' for: '$(Ex) \cdot x\varepsilon a$'.
Explanation: '$\exists!a$' signifies that the class a is not a null class, i.e. that there is at least one element in a. The existence of the class itself must be distinguished from the existence of elements of the class, even in the case where $\sim \exists!a$, i.e. where $a = \Lambda$, the class a exists even though it is empty.

15.5. *The meaning of the word 'is'*

15.51. The English word 'is' (and the corresponding words in other European languages) has two groups of very different meanings: existential and copulative.
15.52. There are, among others, two existential meanings of the word 'is' (both of which are defined by means of the existential quantifier, '(Ex)', 11.22):
15.521. The existence of a described object ('E!', 14.23).
15.522. The non-emptiness of a class ('$\exists$!', 15.43).
15.53. There are, among others, four copulative meanings of the word 'is':
15.531. The associating of a predicate with an individual ('φa', 11.14).
15.532. An element's belonging to a class ('ε', 15.13), which is defined

by means of a matrix ('φx', 11.15 sq.).
15.533. The inclusion of one class in another ('$\subset$', 15.25).
15.534. Identity ('$=$', 14.11).

15.6. *The unit and dual classes*

15.61. '$[x]$' for: '$\hat{y}(y = x)$' *Unit class.*
Explanation: The class $[x]$ is the class which has only one x for element; e.g. the unit class of terrestrial moons. In spite of this, the class must be distinguished from its element, since it possesses properties which the element does not, such as that of containing an element.
15.62. '$[x, y]$' for: '$[x] \cup [y]$', *Dual class.*
15.63. '1' for: $\hat{a}\{(Ex) \cdot a = [x]\}$ *Cardinal number 1.*
Explanation: The cardinal number one is the class of all unit classes. When I say that I have one pencil, I qualify, not the pencil, but the class of my pencils; this is particularly clear when one goes on to higher numbers: a number is attributed only to a class (Frege).
15.64. '2' for: '$\hat{a}\{(Ex, y) : a = [x, y] \cdot x \neq y\}$'.
Explanation: 2 is the class of all dual classes whose elements are not identical with each other.

HISTORY: The Syllogistic of Aristotle can be interpreted as a logic of classes, although it seems to be largely arbitrary whether a distinction between classes and predicates can be attributed to him. The same holds for the Scholastics. The real creator of the logic of classes was Boole. His 'algebra of logic' was the first part of mathematical logic to be fully elaborated. He used the signs ('$\times$', '$+$', etc.) and operations similar to those of mathematics. Frege, and after him Peano, defined the class on the basis of the sentential and predicate calculus.

LITERATURE: PM *20, *22, *24; for the algebra of logic, Lewis 1; Schröder; modern elaborations: Moisil 1; another system: Leśniewski 1.

§ 16. THE CALCULUS OF CLASSES

16.1. *Analogous laws*

16.11. 'X is an analogous expression (16.1) of Y' for: 'X is an expression formed by substituting '$x\varepsilon\alpha$' for 'φx', '$x\varepsilon\beta$' for 'ψx', '$x\varepsilon\gamma$' for 'χx' in Y, in which a point is added to each group of points'.
Example: '$(x) : x\varepsilon\alpha \cdot \equiv \cdot x\varepsilon\alpha$' is an analogous expression (16.1) for '$(x) \cdot \varphi x \equiv \varphi x$' (12.53)'.
16.12. Every analogous expression (16.1) of a law in § 12 or obtained by virtue of the rules of § 12.5 is itself a law.

16.2. *Principal laws*

16.211. $\alpha \cup \beta \cdot = \cdot \beta \cup \alpha$ (5.22)
16.212. $\alpha \cap \beta \cdot = \cdot \beta \cap \alpha$ (5.52)
16.221. $\alpha \cdot \cup \cdot \beta \cup \gamma : =: \alpha \cup \beta \cdot \cup \gamma$ (5.23)
16.222. $\alpha \cdot \cap \cdot \beta \cap \gamma : = : \alpha \cap \beta \cdot \cap \gamma$ (5.53)
16.231. $\alpha \cup \alpha \cdot = \alpha$ (5.15)
16.232. $\alpha \cap \alpha \cdot = \alpha$ (5.16)
16.241. $\alpha = \alpha$ (5.11)
16.242. $\alpha \subset \alpha$ (5.11)
16.243. $- - \alpha = \alpha$ (5.12)
16.25. $\alpha \subset \beta \equiv - \beta \subset - \alpha$ (5.32)
16.26. $\alpha \subset \beta : \supset : \beta \subset \gamma \cdot \supset \cdot \alpha \subset \gamma$ (6.32)
16.27. $\alpha \subset \beta \cdot \equiv \cdot \alpha \cap \beta = \cdot \alpha$ (5.314)
16.28. $\alpha \subset \beta \cdot \equiv \cdot \alpha \cup \beta = \cdot \beta$ (5.315)
16.29. $\alpha \subset \beta \cdot x\varepsilon\alpha \cdot \supset x\varepsilon\beta$ (12.91)

16.3. *Laws of the universal and the null class*

16.311. $\Lambda = - \mathbf{V}$
16.312. $\Lambda \neq \mathbf{V}$
16.313. $\mathbf{V} = - \Lambda$
16.321. $(x) \cdot x\varepsilon \mathbf{V}$
16.322. $(x) \cdot x \sim \varepsilon \Lambda$
16.323. $(\alpha) \cdot \alpha \subset \mathbf{V}$
16.324. $(\alpha) \cdot \Lambda \subset \alpha$
16.331. $(x) \cdot x\varepsilon\alpha \cdot \equiv \cdot \alpha = \mathbf{V}$
16.332. $(x) \cdot x \sim \varepsilon\alpha \cdot \equiv \cdot \alpha = \Lambda$

16.341. $\alpha = \mathbf{V} \cdot \equiv \cdot - \alpha = \Lambda$
16.342. $\alpha \cup - \alpha = \mathbf{V}$
16.343. $\alpha \cap - \alpha = \Lambda$
16.344. $\alpha \cup \Lambda = \alpha$
16.345. $\alpha \cap \Lambda = \Lambda$
16.346. $\alpha \cup \mathbf{V} = \mathbf{V}$
16.347. $\alpha \cap \mathbf{V} = \alpha$
16.351. $\alpha \subset \beta \cdot \equiv \cdot - \alpha \cup \beta = \mathbf{V}$
16.352. $\alpha \subset \beta \cdot \equiv \cdot \alpha \cap - \beta = \Lambda$
16.353. $- \alpha \subset \beta \cdot \equiv \alpha \cup \beta \cdot = \cdot \mathbf{V}$
16.354. $\alpha \subset - \beta \cdot \equiv \cdot \alpha \cap \beta \cdot = \cdot \Lambda$
16.361. $\alpha \cup \beta \cdot = \cdot \Lambda : \equiv : \alpha = \Lambda \cdot \beta = \Lambda$
16.362. $\alpha \cap \beta \cdot = \cdot \Lambda : \equiv : \alpha \cdot = \cdot \alpha \cap - \beta$
16.363. $\alpha \cap \beta \cdot = \cdot \Lambda : \equiv : (x, y) : x \varepsilon \alpha \cdot y \varepsilon \beta \cdot \supset \cdot x \neq y$
16.371. $\alpha : = : \alpha \cap \beta \cdot \cup \cdot \alpha \cap - \beta$
16.372. $\beta \subset \alpha \cdot : \supset : \cdot \alpha : = : \beta \cdot \cup \cdot \alpha \cap - \beta$

16.4. *Laws of existence*

16.411. $\sim \exists ! \alpha \cdot \equiv \cdot \alpha \cdot = \Lambda$
16.412. $\exists ! \alpha \cdot \equiv \cdot \alpha \neq \Lambda$
16.421. $\exists ! \mathbf{V}$
16.422. $\sim \exists ! \Lambda$
16.431. $\exists !(\alpha \cup \beta) : \equiv : \exists ! \alpha \cdot \vee \cdot \exists ! \beta$
16.432. $\exists !(\alpha \cap \beta) : \supset : \exists ! \alpha \cdot \exists ! \beta$
16.433. $\alpha \subset \beta : \supset : \exists ! \alpha \cdot \supset \cdot \exists ! \beta$
The inverse of the two above is not true
16.44. $\sim \cdot \alpha \subset \beta : \equiv : \exists !(\alpha \cap - \beta)$
16.451. $\alpha \cap \beta \cdot = \cdot \Lambda : \supset : \exists ! \alpha \cdot \supset \cdot \alpha \neq \beta$
16.452. $\exists ! \alpha \cdot \alpha = \beta \cdot \supset \cdot \exists !(\alpha \cap \beta)$
16.453. $\alpha \subset \beta \cdot \alpha \neq \beta \cdot \equiv \cdot \exists !(- \alpha \cap \beta)$
16.461. $\sim \exists ! \beta \cdot \supset \cdot \alpha \cup \beta \cdot = \cdot \alpha$
16.462. $\sim \exists ! \beta \cdot \supset \cdot \alpha \cap \beta \cdot = \cdot \Lambda$

HISTORY and LITERATURE: See § 15.

§ 17. ANTINOMIES AND THE THEORY OF TYPES

This chapter presents a short, elementary, and non-formalized exposition of the antinomies (also called 'paradoxes') which arise in logical systems and of the rules for avoiding them. One group of these rules is known as the 'theory of types'.

17.1. *Antinomies*

17.11. '*Antinomy*' for: 'the logical product of a sentence and the negation of its equiform or an equivalent expression'.
Examples: '$p \cdot \sim p$', '$(x)\varphi x \cdot (Ex) \sim \varphi x$' are antinomies.
17.12. In any sufficiently formalized logical system an indefinite number of antinomies can be deduced, if special precautions are not observed.
17.13. The antinomies can be divided into logical and metalogical or semantic antinomies.
17.14. '*Logical antinomy*' for: 'antinomy which arises within the logical system itself without any use of metalogical expressions'.
17.15. '*Metalogical*' or '*Semantic antinomy*' for: 'antinomy which arises from the use of metalogical expressions'.
Examples: 17.14: antinomy of the class of classes (cf. 17.2); 17.15: antinomy of the liar (cf. 17.7).

17.2. *The antinomy of the class of classes*

17.21. We form the class of all classes which do not contain themselves and then pose the question, whether this class contains itself. From the affirmative answer we can deduce that it does not contain itself and from the negative answer that it does contain itself. This antinomy is called after its discoverer, Russell's Paradox.
Justification: If '$\alpha \varepsilon \alpha$' is an expression, we can define (15.11) a class β such that, for all α

(1) $\alpha \varepsilon \beta \cdot \equiv \cdot \sim \cdot \alpha \varepsilon \alpha$

Substituting 'β' for 'α', we obtain

(2) $\beta \varepsilon \beta \cdot \equiv \cdot \sim \cdot \beta \varepsilon \beta$

and from this we get

(3) $\beta \varepsilon \beta \cdot \sim \cdot \beta \varepsilon \beta$

which is an antinomy (17.11).

Example: A library catalogue provides a record of the books in the library. The catalogue itself can be considered a book and catalogued accordingly. If now we were to draw up a complete catalogue of all the catalogues which do not include themselves, the question would arise whether we should include the catalogue we were making. If we do, then we no longer have a catalogue which does not include itself, and it must be excluded. But if we exclude it, then we have a catalogue which does not contain itself, and it qualifies for inclusion In either case we derive the opposite of our assumed condition.

17.22. The expression '$\alpha \varepsilon \alpha$' is meaningless.

Proof: If it were meaningful, (1) would be true or not true; it could not be both. It appears to be a sentence, but is not. It is a group of signs signifying nothing.

17.3. *The Theory of Types*

17.31. '*Theory of Types*' for: 'the set of rules which, by dividing objects or logical expressions into numbered classes (types), makes it possible to avoid the logical antinomies'.

17.32. '*Theory of ontological types*' for: 'the theory of types which divides *objects* into types'.

Explanation: A theory of ontological types has as 1st type the set of individuals, as 2nd that of classes of individuals, as 3rd that of classes of classes of individuals, etc.

17.33. '*Theory of syntactical types*' for: 'the theory of types which divides *expressions* into types'.

Example: 17.4.

17.4. *Rules of syntactical types*

17.41. *Rule:* All expressions are divided into numbered classes which are mutually exclusive, called 'Type 1', 'Type 2', ..., 'Type n'. These 'Types' are another way of dividing the syntactical categories (1.22).

17.42. *Rule:* All equiform expressions of the same system belong to the same type.

17.43. *Rule:* If F is a functor of X and X belongs to type n, F belongs to type $n + 1$.

17.44. *Rule:* If X is followed by 'ε' followed by Y, and X belongs to type n, then Y belongs to type $n + 1$.

17.5. *Quine's Method of Verification*

17.51. To verify whether rule 17.44 has been followed, one may proceed as follows:
(a) for all equiform variables arbitrarily chosen substitute '*O*'; (b) if a variable is immediately followed by 'ε' followed immediately by a numeral, substitute for this variable a numeral smaller by one (positive, '*O*', negative); (c) repeat this operation until all the variables are replaced by numerals; if necessary, start over with a new variable; (d) if then the numeral which immediately follows each 'ε' is always larger by one than that which preceeds it, the expression conforms to the rule of types; if not, it does not.
17.52. To verify whether rule 17.43 has been followed, first substitute for the expressions 'φx', 'ψx', 'χx' respectively '$x\varepsilon\alpha$' '$x\varepsilon\beta$', '$x\varepsilon\gamma$', and then apply 17.51.

17.6. *The Principle of Analogy*

17.61. The application of the rule of types makes it necessary to distinguish as different types of constant functors 'ε', '=', '≠', etc. and the expressions 'V', '∧'.
Explanation: The rule of types applies also to dyadic functors such as '='. Consequently, if '=' in '$x = x$' belongs to type *n*, and if we also have '$x\varepsilon\alpha$', we cannot according to the rule of types write '$\alpha = \alpha$' since 'α' being of a higher type than *x*, the '=' which unites the two 'α' must also be of a higher type than the first '=', which is opposed to rule 17.42.
17.62. To avoid the multiplication of expressions and laws for each type rule 17.42 is not applied to the functors enumerated in 17.61.
17.63. *Principle of analogy or systematic ambiguity:* The functors 'ε', '=', '≠', etc. and the expressions 'V' and '∧' are systematically ambiguous with respect to type.
Explanation: Expressions of this form have a different meaning according to their type, but their formal properties remain the same; e.g. laws 14.13–15 remain valid when the individual names (variables) that they contain are replaced by class names.

17.7. *The Antinomy of the Liar*

17.71. '*Antinomy of the liar*' for: 'the antinomy which results from introducing into the system expressions of the type '*X* is false''.
17.72. In any system of formalized logic which contains the laws and rules

given in the preceeding chapters it is possible to deduce the antinomy of the liar by introducing '*X* is false', unless special precautions are observed. Justification: Form the sentence '*c* is false' and take '*c*' as the typographical abbreviation of this sentence. We then have

(1) c is false $= c$

However, according to the current definition of the truth of a sentence, we have

(2) X is true $\cdot \equiv \cdot X$

Substituting '*c* is false' for '*X*' in (2), we obtain

(3) c is false is true $\cdot \equiv \cdot$ c is false

Substituting, from the identity (1), '*c*' for '*c* is false', we obtain from (3)

(4) c is true $\cdot \equiv \cdot$ c is false

Using the definition of falsity

(5) c is false $\cdot \equiv \cdot \sim \cdot$ c is true

we obtain from (4) the antinomy

(6) c is true $: \equiv : \sim \cdot$ c is true

17.8. *Solution of the metalogical antinomies*

17.81. *Rule:* To avoid the metalogical antinomies it is necessary to observe strictly the rules of supposition (2.13 or 2.14).

17.82. If rule 2.13 is followed, the antinomy of the liar does not appear. Justification: In this case instead of (1) and (2) in 17.72 we obtain respectively

(1′) c is false $= c$

(2′) 'X' is true $\cdot \equiv \cdot X$.

But we can proceed no further, since the '*X*' at the beginning of (2′) is the name of the latter and nothing can be substituted for it.

A fully satisfactory solution of the antinomy of the liar demands an elaboration of the definition of truth.

HISTORY: The antinomies were known in antiquity; and were rediscovered and thoroughly studied by the Scholastics. Around 1900 the paradoxes of set theory shook the foundations of mathematics. In 1908 Russell and Zermelo both offered different solutions of the problem. The 'simple theory of types' of Russell, taken up in PM, was developed into a 'ramified theory' (PM 1st edit.). Later (PM 2nd edit.) a tendency towards simplification appears and a laying aside of the restrictions imposed by the theory of types. Some, like Ushenko, think that it is possible to do without a general theory of types. Type-free systems of logic have been developed by Bernays and Ackermann.

– The distinction between the logical and semantic paradoxes comes from Ramsey.

LITERATURE: Russell 3, Russell 4, PM (Introd. 2nd edit.); Zermelo; Chwistek 1; Ramsey; Tarski 2; Quine 2, Quine 3; Church 2; Fraenkel B; Fitch 1; Ackermann 1; Bernays 1; Behmann 1; Ushenko 1.
– History: Rüstow; Salamucha.

IV

THE LOGIC OF RELATIONS

§ 18. RELATIONS

The calculus of relations bears the same relation to the theory of dyadic predicates (§ 13) that the calculus of classes (§ 15) does to the theory of monadic predicates (§ 12). It is the newest and also the most important part of modern logic. Developed originally for the foundations of mathematics, it has gone beyond this science to embrace the whole of knowledge. Despite the fact that it occupies a major place in the treatises of logic, it is still relatively little developed.

18.1. *Definitions*

18.11. '$\hat{x}\hat{y}\{\varphi x, y)\}$' for: 'the x's and y's such that $\varphi(x, y)$'.
Explanation: Cf. 15.11; the couples so defined are called 'relations'. Thus 'relation' is here taken in extension.
18.12. '*Rel*' for: '$\hat{R}\{(E\varphi) \cdot R = \hat{x}\hat{y}[\varphi(xy)]\}$'.
Explanation: Cf. 15.12. Rel is the class of relations, i.e. of the couples defined in 18.11.
18.13. '$u\{\hat{x}\hat{y}[\varphi(xy)]\}v$' for: '$\varphi(u, v)$'.
18.14. 'uRv' for: '$u\{\hat{x}\hat{y}[\varphi(x, y)]\}v$'.
Explanation: 18.13–14 serve to introduce the new notation 'xRy'.
18.15. 'Antecedent of R' for: 'The object which has the relation R to something'.
18.16. 'Consequent of R' for: 'the object to which something has the relation R'.
18.17. 'Term of R' for: 'antecedent or consequent of R'.

18.2. *Relations between relations*

18.21. '$\dot{-}R$' for: '$\hat{x}\hat{y}(\sim xRy)$'.
Explanation: $\dot{-}R$ is the complementary relation of R (cf. 15.21), i.e. the class of all couples not joined by the relation R. Example: the complementary relation of 'brother' is the set of couples who are friends, neighbors, larger than, superior to, similar to, etc., but are not brothers.

18.22. '$R \,\dot{\cup}\, S$' for: '$\hat{x}\hat{y}(xRy \vee xSy)$' (15.22)
18.23. '$R \,\dot{\cap}\, S$' for: '$\hat{x}\hat{y}(xRy \cdot xSy)$' (15.23)
18.24. '$R \,\dot{\|}\, S$' for: '$\hat{x}\hat{y}(xRy \mid xSy)$' (15.24)
18.25. '$R \,\dot{\subset}\, S$' for: '$(x, y) \cdot xRy \supset xSy$' (15.25)
18.26. '$R \doteq S$' for: '$(x, y) \cdot xRy \equiv xSy$' (15.26)
18.27. '$\dot{\mathbf{V}}$' for: '$\hat{x}\hat{y}(x = x \cdot y = y)$' (15.41)
18.28. '$\dot{\Lambda}$' for: '$\hat{x}\hat{y}(x \neq x \cdot y \neq y)$' (15.42)
18.29. '$\dot{\exists}!R$' for: '$(Exy)xRy$' (15.43)

The names of these functors are the same as for classes: 'sum of relations', 'null relation', etc.

18.3. *Analogous Laws*

18.31. 'X is an analogous expression (18.3) of Y' for: 'X is an expression formed by substituting for
'α', 'β', 'γ', 'δ', '$-$', '$\cap$', '$\cup$', '$\|$', '$\supset$', '$=$', '$\vee$', '$\wedge$',
respectively
'P', 'Q', 'R', 'S', '$\dot{-}$', '$\dot{\cap}$', '$\dot{\cup}$', '$\dot{\|}$', '$\dot{\subset}$', '$\doteq$', '$\dot{\mathbf{V}}$', '$\dot{\Lambda}$', in Y'.

18.32. Every analogous expression (18.3) of a law in the class calculus (§ 16, including the laws formed by 16.11) is itself a law.

HISTORY: The elements of a theory of relations are found first in the *Topics* of Aristotle, but its full development does not come until the 19th century. The idea of defining relation as a class of couples comes from Peirce; it was further developed by Frege and Peano. Its present form is due to PM. Wiener and Kuratowski have provided a new basis for the theory.

LITERATURE: PM *21, *23, *25, and other textbooks. For new developments: Wiener; Kuratowski; Tarski 5; Quine 3, § 36. An exceptionally clear elementary exposition is in Carnap 1, 8.

§ 19. RELATIVE DESCRIPTIONS; CONVERSE

Descriptions are of particular importance when they concern dyadic predicates (relations). Most current expressions, for example in theology, law, and mathematics, are, in fact, relative descriptions. Several different kinds of these descriptions are known. – At the end of the section another important theory is given, that of the converse of a relation.

19.1. *Individual and Plural Descriptions*

19.11. ‘$R'y$’ for: ‘$(\iota x)(xRy)$’.

Explanation: Read ‘R of y’. This expression is called the ‘individual relative description’, since it describes *only one* individual which has a given relation to one other object (cf. 14.22). Example: if ‘R’ signifies ‘author of’ and ‘a’ the ‘*Iliad*’, ‘$R'a$’ signifies: ‘the author of the *Iliad*’. This expression will be meaningless on the theory of Wolff that the *Iliad* had several authors.

19.12. ‘$\overrightarrow{R}'y$’ for: ‘$\hat{x}(xRy)$’.

19.13. ‘$\overleftarrow{R}'x$’ for: ‘$\hat{y}(xRy)$’.

Explanation: These two expressions are called ‘plural relative descriptions’, since they signify the *class* of objects which have the relation R to a given individual (‘$\overrightarrow{R}'y$’) or to which a given individual has the relation R (‘$\overleftarrow{R}'x$’). Example: If ‘R’ signifies ‘author of’ and ‘a’ the ‘Bible’, ‘$\overrightarrow{R}'a$’ is ‘the class of authors of the Bible’. If ‘a’ signifies ‘Homer’, ‘$\overleftarrow{R}'a$’ signifies ‘the works (the class of works’) of Homer’.

Graphical representation $\qquad \overleftarrow{R}'x \qquad \overrightarrow{R}'y$

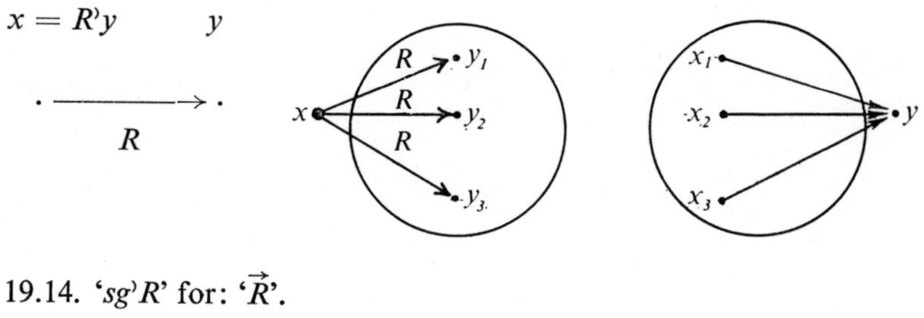

19.14. ‘$sg'R$’ for: ‘$\overrightarrow{R}$’.

19.15. ‘$gs'R$’ for: ‘$\overleftarrow{R}$’.

Explanation: 'sg'R' (from *sagitta*) and 'gs'R' (*sagitta* backwards) serve to replace '$\overrightarrow{R}$' and '$\overleftarrow{R}$' in longer formulas, e.g. sums, products, of relations, etc.

19.2. *Bi-plural Descriptions*

19.21. '$R``\beta$' for: '$\hat{x}\{(Ey) \cdot y\varepsilon\beta \cdot xRy\}$'.
Explanation: $R``\beta$ is the class of individuals which have the relation $R$ to any element of the class $\beta$. Example: If '$R$' is 'author of' and $\beta$ is the class of writings in verse, $R``\beta$ is the class of authors of writings in verse, or as some think wrongly, of poets.
Graphical representation:

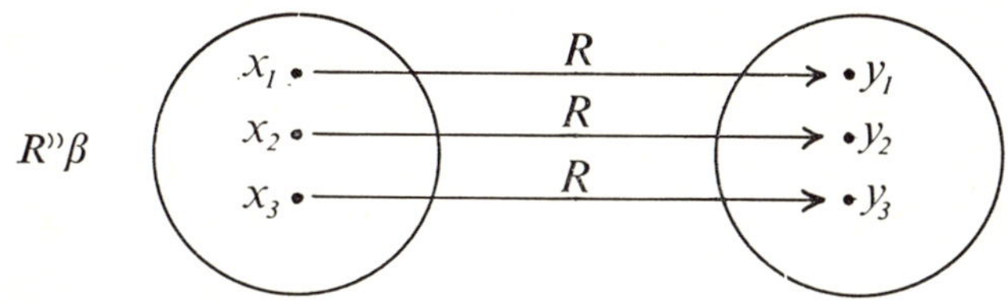

19.22. (α, β, R): $\alpha \subset \beta \cdot \supset \cdot R``\alpha \subset R``\beta$.
Examples: If horses are animals, the heads of horses are heads of animals.

19.3. *Converse*

19.31. '$x\breve{R}y$' for: 'yRx'.
Example: If 'R' is 'author of', '$\breve{R}$' is 'the work of'. If 'R' is 'to the right of', '$\breve{R}$' is 'to the left of'.
19.32. 'Cnv'R' for: '$\breve{R}$'.
Explanation: Cnv is a relation which holds between R and $\breve{R}$. The description 'Cnv'R' serves to replace '$\breve{R}$' when a longer expression is substituted for 'R'.
Graphical representation:

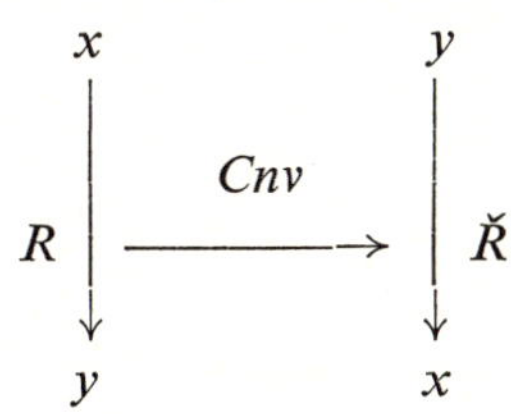

19.4. *Laws of the Converse*

19.41. $R \doteq S \cdot \equiv \cdot \breve{R} \doteq \breve{S}$

19.42. $Cn'vCnv'R \doteq R$
19.43. $\exists ! Cnv'R$
19.44. $Cnv'(R \mathbin{\dot\cup} S) \cdot \doteq \cdot Cnv'R \mathbin{\dot\cup} Cnv'S$
19.45. $Cnv'(R \mathbin{\dot\cap} S) \cdot \doteq \cdot Cnv'R \mathbin{\dot\cap} Cnv'S$
19.46. $Cnv' \dot- R \doteq \dot- Cnv'R$
19.47. $R \doteq \breve{S} \cdot \equiv \cdot S \doteq \breve{R}$
19.48. $R \mathbin{\dot\subset} \breve{S} \cdot \equiv \cdot S \mathbin{\dot\subset} \breve{R}$

HISTORY: The theory of relative descriptions was considered as early as the time of De Morgan, who gave 19.22. The theory of the converse was elaborated by mathematicians in connection with the theory of sets. It was examined by Cayley in 1854. Frege and Peano, the founders of contemporary logic, and the authors of PM provided its present form.

LITERATURE: PM *30 – *32 and all the textbooks.

§ 20. DOMAINS AND FIELDS

Besides the expressions given in § 19, there are also similar but more general notions; we not only use expressions such as 'the mother of', 'the neighbor of', but also the more general expressions, 'the mothers', 'the neighbors'. The theory of such expressions is given in this chapter, followed by certain information regarding relations that are limited from one or two sides to a unit class.

20.1. *Domains and Fields*

20.11. '$D^{\prime}R$' for: '$\hat{x}\{(Ey)xRy\}$'.

Explanation: $D^{\prime}R$ is the *domain* of R, i.e. the set of objects which stand in the relation R to any object. Example: If 'R' signifies 'father of', '$D^{\prime}R$' is the class of all fathers.

20.12. '$\text{Я}^{\prime}R$' for: '$\hat{y}\{(Ex)xRy\}$'.

Explanation: $\text{Я}^{\prime}R$ is the *converse domain* of R, i.e. the set of objects to which any other objects stand in the relation R.

Example: If 'R' is 'husband of', '$\text{Я}^{\prime}R$' is the class of all wives.

20.13. '$C^{\prime}R$' for: '$D^{\prime}R \cup \text{Я}^{\prime}R$'.

Explanation: $C^{\prime}R$ is the *field* (from *campus*) of R, i.e. the logical sum of the domain and the converse-domain of R. Examples: If 'R' is 'military superior of', '$D^{\prime}R$' is the class of all those who are military superiors, i.e. all the officers of all armies; '$\text{Я}^{\prime}R$' is the class of all those who have military superiors, i.e. of all soldiers save the commanders-in-chief; finally '$C^{\prime}R$' is the class consisting of both. If 'R' is 'parent of', '$D^{\prime}R$' is the class of all men and women who have children; '$\text{Я}^{\prime}R$' is the class of all those who have parents, i.e. of all human beings except Adam and Eve.

The difference between $D^{\prime}R$ and $\vec{R}^{\prime}y$ and $R^{\prime\prime}a$ consists in this that $\vec{R}^{\prime}y$ is the class of objects which have the relation R to a definite individual y, and $R^{\prime\prime}a$ that of objects which have this relation to the elements of a definite class a, whereas $D^{\prime}R$ is the class of all objects having the relation R to any object.

20.2. *Laws of Domains and Fields*

20.21. $(x, y) : xRy \cdot \supset \cdot x\varepsilon D^{\prime}R \cdot y\varepsilon \text{Я}^{\prime}R$

20.22. $(y) \cdot \vec{R}^{\prime}y \subset D^{\prime}R$

20.23. $(x) \cdot \overleftarrow{R}\text{'}x \subset \text{ᗡ}\text{'}R$
20.24. $D\text{'}R = \text{ᗡ}\text{'}\breve{R}$
20.25. $C\text{'}R = C\text{'}\breve{R}$
20.26. $C\text{'}R = C\text{'}(R \,\dot{\cup}\, \breve{R})$
20.27. $D\text{'}R \subset \text{ᗡ}\text{'}R \cdot \equiv \cdot \text{ᗡ}\text{'}R = C\text{'}R$
Explanation: If the domain of R is included in the converse domain of R, the latter is equal to the field of R. In this case the series formed by R has no beginning, since for any term of R there is always an element of $\text{ᗡ}\text{'}R$, i.e. there is an antecedent.
20.28. $\text{ᗡ}\text{'}R \subset D\text{'}R \cdot \equiv \cdot D\text{'}R = C\text{'}R$
Explanation: In this case the series has no end, since for any term of R there is always an element of $D\text{'}R$, i.e. there is a consequent.

20.3. *Relations with Limited Domains*

20.31. '$\alpha \rceil R$' for: '$\hat{x}\hat{y}(x\varepsilon\alpha \cdot xRy)$'
20.32. '$R \upharpoonright \beta$' for: '$\hat{x}\hat{y}(y\varepsilon\beta \cdot xRy)$'
20.33. '$\alpha \rceil R \upharpoonright \beta$' for: '$\hat{x}\hat{y}(x\varepsilon\alpha \cdot y\varepsilon\beta \cdot xRy)$'
20.34. '$R \upharpoonleft \alpha$' for: '$\alpha \rceil R \upharpoonright \alpha$'
Explanation: 20.31–34 introduce the notion of a relation with limited domains and fields. Thus $\alpha \rceil R$ is the relation R limited in its domain to the class α, $R \upharpoonright \beta$ the same relation limited in its converse domain to the class β, $\alpha \rceil R \upharpoonright \beta$ the relation R limited in its domain to the class α and in its converse domain to the class β; finally, $R \upharpoonleft \alpha$ is the relation R whose field is limited the class α. Example: If 'R' is 'author of' and 'α' is 'Italian', $\alpha \rceil R$ is the relation of author restricted in its domain to Italians; in this case, $D\text{'}(\alpha \rceil R)$ is the class of Italian authors and $\text{ᗡ}\text{'}(R \upharpoonright \alpha)$ that of Italian works.
20.35. '$\alpha \uparrow \beta$' for: '$\hat{x}\hat{y}(x\varepsilon\alpha \cdot y\varepsilon\beta)$'
20.36. $\alpha \uparrow \beta = \alpha \rceil \dot{V} \upharpoonright \beta$
Explanation: $\alpha \uparrow \beta$ is the relation which exists between x and y by the very fact that x is an element of α and y an element of β; it is the meaning that '$\alpha \rceil R \upharpoonright \beta$' obtains if '$\dot{V}$' is substituted for '$R$'. This notion plays an important role in the theory of series.
20.37. '$x \downarrow y$' for: '$[x] \uparrow [y]$'
Explanation: 20.37 gives the definition of the ordinal couple.

20.4. *One-One Relations*

20.41. '$1 \rightarrow Cls$' for: '$\hat{R}\{(x, y, z): xRz \cdot yRz \cdot \supset \cdot x = y\}$'
20.42. '$Cls \rightarrow 1$' for: '$\hat{R}\{(x, y, z): xRy \cdot xRz \cdot \supset \cdot y = z\}$'
20.43. '$1 \rightarrow 1$' for: '$(1 \rightarrow Cls) \cap (Cls \rightarrow 1)$'
Explanation: 20.41 defines the *one-many relation*, i.e. restricted in its domain to unit classes; 20.42 defines the *many-one relation*, which is restricted in the same way in its converse-domain; 20.43 defines the *one-one relation* in which both domain and converse-domain are restricted to unit classes. 20.41 says in effect that whenever xRz and yRz occur there is identity between x and y so that there can never be but one antecedent for R; 20.42 says the same thing of the consequent. Examples: The relation 'father' is one-many, for the same father can have several children, but a child can have only one father. For Mohammedans the relation 'husband of' would be one-many and 'wife' many-one, but for Christians both relations are one-one.
20.44. $R\varepsilon(1 \rightarrow Cls) \cdot \equiv \cdot \breve{R}\,\varepsilon(Cls \rightarrow 1)$

LITERATURE: § 20.1–2: PM *33; § 20.3; PM *35; § 20.4; PM *71; Carnap 1, 8.

§ 21. RELATIVE PRODUCT; SERIES

The notion of a relative product is important for all sciences which, like mathematics and theology, use the concept of series. This chapter presents the fundamental notions and some elementary applications of the theory of series. An analysis of series by itself would form an extensive treatise.

21.1. *Relative Product*

21.11. 'R/S' for: '$\hat{x}\hat{z}\{(Ey) \cdot xRy \cdot ySz\}$'
Explanation: R/S is the *relative product* of R and S, i.e. the relation which exists between x and z if there is a y such that xRy and ySz. Example: If 'R' is 'father' and 'S' is 'brother', 'R/S' is 'uncle', namely the y such that 'x is the father of y' and 'y is the brother of z'. The relative product of the square and the half is the square of the half.
21.12. 'R^2' for: 'R/R'
21.13. 'R^3' for: 'R^2/R'
21.14. 'R^n' for: 'R^{n-1}/R'
21.15. 'R^0' for: '$I \upharpoonright C\text{'}R$'
Explanation: Expressions 21.12–15 are called 'relative powers' ('relative square', 'relative cube', etc.). R^0 is identity (cf. 14.16) restricted to the field of R, i.e. the relation of identity that each element of $C\text{'}R$ has to itself; this notion has a role in series similar to that of zero in mathematics. Examples: If 'R' signifies 'father of', 'R^2' is 'paternal grandfather of'. The adage 'the friends of my friends are my friends' would translate as '$R^2 \subset R$', where 'R' is 'friend of'.

21.2. *Ancestral Relation*

21.21. '*her*' for: '$\hat{a}(\exists\breve{R})(R\text{''}a \subset a)$'
Explanation: A class is called 'hereditary' with respect to the relation R (*her* is the class of hereditary classes), when the consequents of R in relation to the elements of a are elements of a.
Examples: The class of Hungarians is hereditary with respect to the relation of father, for if x is the father of y and x belongs to the class of Hungarians, i.e. if he is a Hungarian, then y is also a Hungarian.
21.22. 'R_*' for: '$\hat{x}\hat{y}\{x \varepsilon C\text{'}R : \cdot (a) : \breve{R}\text{''}a \subset a \cdot x \varepsilon a \cdot \supset \cdot y \varepsilon a\}$'
Explanation: 21.22 is an ingenious definition of the vague notion '$R^0 \cup R \cup R^2 \cup R^3$ etc.', and hence of the relation which exists when

some power of R is given. It is called the 'ancestral relation'. Examples: If 'R' is 'father of', 'R_*' is 'paternal ancestor'; if 'R' is 'immediately to the left of', 'R_*' is 'to the left of (at any distance)'; if 'R' is 'immediate superior', 'R_*' is 'superior (immediate or otherwise)'.

21.23. 'R_{po}' for: '$\hat{x}\hat{y}\{(\alpha): \breve{R}``\alpha \supset \alpha \cdot \overleftarrow{R}`x \varepsilon \alpha \cdot \supset \cdot y \varepsilon \alpha\}$'
Explanation: R_{po} is distinguished from R_* by the fact that it excludes R^0. It is the equivalent of '$R \cup R^2 \cup R^3$ etc.'.

21.3. *First and last terms*

21.31. 'B' for: '$\hat{x}\hat{R}\{x \varepsilon \cdot D`R \cap - Ɑ`R\}$'
Explanation: 'B' (from 'beginning') is the relation between the first term x of the series formed by R and R itself; 21.31 says in effect that x belongs to domain but not to the converse-domain of R. The class of first terms of R is $sg`B`R$, and that of the last terms $sg`B`Cnv`R$.
21.32. 'Min_R' for: '$\hat{x}\hat{\alpha}\{x \varepsilon \cdot \alpha \cap C`R \cap - \breve{R}``\alpha\}$'
21.33. 'Max_R' for: '$Min_{\breve{R}}$'
Explanation: Min_R is B restricted to one class; it is the minimum of this class with respect to R. Max_R is the maximum.

21.4. *Isomorphic relations*

21.41. '$R \dagger S$' for: '$R/S/\breve{R}$'
Explanation: The relation $R \dagger S$ holds between x and t when one has $(\exists y, z) xRy \cdot ySz \cdot z\breve{R}t$; or graphically

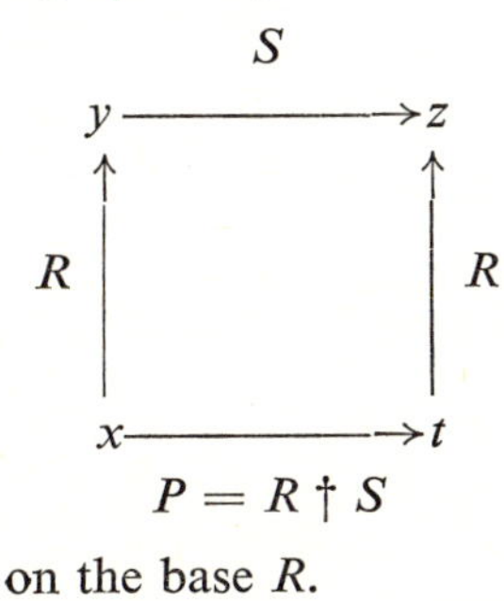

$P = R \dagger S$

$R \dagger S$ is the *image* of S on the base R.
21.42. '$\overline{Psmor}\ S$' for: '$\hat{R}\{R \varepsilon 1 \to 1 \cdot C`S = Ɑ`R \cdot P = R \dagger S$'
21.43. '$Smor$' for: '$\hat{P}\hat{S}\{\exists ! \overline{Psmor\ S}\}$'
Explanation: The relation P is said to be 'isomorphic' ('smor' from the latin 'similis ordine') to S when there exists at least one one-one relation

R such that $P = R \dagger S$. (The isomorphy of relations should not be confused with that of terms; cf. 1.15).
Examples: The relation holding between the fathers of two school-friends is isomorphic to that which exists between the boys, if they are only sons.

LITERATURE: § 21.1: PM *34; § 21.2; PM *90, *91; § 21.3: PM *93; § 21.4: PM *150, *151; Carnap 1, 8.

§ 22. PROPERTIES OF RELATIONS

This chapter provides elementary definitions of certain properties common to large groups of relations, such as reflexivity, transitivity, connexity, etc. This theory is of great importance in the higher branches of logic and mathematics and has numerous applications in other fields.

22.1. *Reflexivity*

22.11. '*refl*' for: '$\hat{R}(R^0 \subset \cdot R)$'
22.12. '*irr*' for: '$\hat{R}(R^0 \subset \cdot \dot{-} R)$'
22.13. $R \,\varepsilon\, refl \cdot \equiv \cdot (x) : x \varepsilon C'R \cdot \supset \cdot xRx$
22.14. $R \,\varepsilon\, irr \cdot \equiv \cdot (x) \sim xRx$

Explanation: *refl* is the class of *reflexive* relations, i.e. such that if an x belongs to their field, these relations hold between x and x. On the other hand, *irr* is the class of *irreflexive* relations.

Examples: Identity and love (according to Aristotle) are reflexive relations, since, according to this philosopher every being is identical with itself and loves itself. On the other hand, the relations of being the father of, greater than, neighbor of, etc. are irreflexive. Note that there are relations which are neither reflexive nor irreflexive, e.g. that of cooking food for.

22.2 *Symmetry*

22.21. '*sym*' for: '$\hat{R}(\breve{R} \doteq R)$'
22.22. '*as*' for: '$\hat{R}(\breve{R} \doteq \dot{-} R)$'
22.23. $R \,\varepsilon\, sym \cdot \equiv \cdot (x) \cdot xRy \equiv yRx$
22.24. $R \,\varepsilon\, as \cdot \cdot (x) \cdot xRy \equiv \sim yRx$

Explanation: *sym* is the class of *symmetrical* relations, *as* that of asymmetrical relations. As in the case of reflexivity, there are relations which are neither one nor the other.

Examples: The relation of being a colleague or a neighbor is symmetrical, whereas the relations of being greater than, smaller than, father, daughter, etc. are asymmetrical.

22.3. *Transitivity*

22.31. '*trans*' for: '$\hat{R}(R^2 \subset \cdot R)$'
22.32. '*intr*' for: '$\hat{R}(R^2 \subset \cdot \dot{-} R)$'
22.33. $R \,\varepsilon\, trans : \equiv : (x, y) : (Ez) \cdot xRz \cdot zRy \cdot \supset \cdot xRy$

22.34. $R \varepsilon\, intr :\equiv: (x, y, z) : xRy \cdot yRz \cdot \supset \cdot \sim xRz$
Explanation: *trans* is the class of *transitive* relations, i.e. those which 'pass over' from one term to another; *intr* is the class of *intransitive* relations. Here again there are relations which are neither one nor the other. Examples: The relations of being to the right of, greater than, smaller than, equal to, identical with, $\supset$, $\equiv$, $=$, $\subset$, are transitive, whereas the relations of being father, son, husband, wife, the square of, etc. are intransitive.
22.35. $trans \cap sym \cdot \subset \cdot refl$
22.36. $as \subset irr$

22.4. *Similarity and Equality*

22.41. '*sim*' for: '*sym* $\cap$ *refl*'
Explanation: *sim* (from the Latin *similis*) is the class of relations of similarity, i.e. of 'nearly the same', as 'nearly equal', 'nearly the same color', etc. All such have the properties of symmetry and reflexivity.
22.42. '*aeq*' for: '*trans* $\cap$ *sym*'
Explanation: *aeq* (from *aequalis*) is the class of relations of equality, i.e. of the same form, color, size, etc.). These relations are transitive and symmetrical.
22.43. $aeq \subset refl$
22.44. $aeq \subset sim$

22.5. *Connexity*

22.51. '*connex*' for: '$\hat{R}(J \upharpoonright C'R \cdot \subset \cdot R \cup \breve{R})$'
Explanation: A relation R is said to be 'connected' or 'connex' when R or $\breve{R}$ always holds between any two different objects belonging to the field of the relation. Example: 'greater than' is connex in the field of numbers, since of any two different numbers one is always greater than the other.
22.52. '*ser*' for: '*irr* $\cap$ *trans* $\cap$ *connex*'
Explanation: A relation forms a 'series' when it is irreflexive, transitive, and connex. This relation is of the greatest importance in mathematics and other sciences.
22.53. $ser \subset irr$
22.54. $ser \subset as$
22.55. $R \varepsilon\, ser \cdot \equiv \cdot \breve{R} \varepsilon\, ser$

LITERATURE: PM *201, *202, *204; Carnap 1.

§ 23. POLYADIC RELATIONS

The theory of dyadic relations (with two arguments), although very important, is not sufficient for even the most elementary analysis in the non-mathematical sciences. Unfortunately, it is the only part of the logic of relations that has been developed. This chapter provides some of the fundamental definitions for a general theory of relations.

23.1. *Fundamental Definitions*

In the following definitions 'n' is a variable for which positive integers are to be substituted.

23.11. '$\hat{x}_1, \ldots \hat{x}_n \; \varphi(x_1, \ldots x_n)$' for: 'the $\hat{x}_1, \ldots \hat{x}_n$ such that $\varphi(x_1, \ldots x_n)$'. Cf. 18.11.

23.12. 'Rel_n' for: '$\hat{R}\{(E\varphi) \cdot R = \hat{x}_1, \ldots \hat{x}_n[\varphi(x_1, \ldots x_n]\}$'. Cf. 18.12.

23.13. '$x_1 R(x_2, \ldots x_n)$' for: '$R(x_1, \ldots x_n)$'

23.14. 'Term of R' for: 'an object entering in any way in the relation R with other terms'.

Explanation: If R has more than 2 terms, one can no longer speak of the antecedents and consequents of R (cf. 18.15–16), but only of the nth term of R. The same holds for the converse domain, etc.

23.15. The relations between relations of more than two terms are analogous to those of dyadic relations (cf. 18.2).

Examples: For a triadic relation R

'$\dot{-} R$' is '$\hat{x}_1, \hat{x}_2, \hat{x}_3\{\sim R(x_1, x_2, x_3)\}$';

for two triadic relations R and S

'$R \dot{\cup} S$' is '$\hat{x}_1, \hat{x}_2, \hat{x}_3\{R(x_1, x_2, x_3) \vee S(x_1, x_2, x_3)\}$'.

The meaning of the functors here is obviously different from that defined in § 18.2, but the principle of analogy (17.6) can be applied without trouble.

23.16. Rule 18.32 applies to relations of more than two terms.

23.2. *Relative descriptions*

23.21. '$R_1{}^{\flat}(x_2, \ldots x_n)$' for: '$(\iota x_1)\{R(x_1, \ldots x_n)\}$'. Cf. 19.11.

23.22. '$R_k{}^{\flat}(x_1, \ldots x_{k-1}, x_{k+1} \ldots x_n)$' for: '$(\iota x_k)\{R(x_1, \ldots x_k, \ldots x_n)\}$'.

23.23. '$\vec{R}_1{}^{\flat}(x_2, \ldots x_n)$' for: '$\hat{x}_1\{R(x_1, \ldots x_n)\}$'. Cf. 19.12.

23.24. '$\vec{R}_k{}^{\flat}(x_1, \ldots x_{k-1}, x_{k+1}, \ldots x_n)$' for: '$\hat{x}_k\{R(x_1, \ldots x_k, \ldots x_n)\}$'.

23.25. '$sg_k{}^{\flat}R$' for: '$\vec{R}_k$'. Cf. 19.14.

The theory of multi-plural descriptions (corresponding to bi-plural descriptions 19.2) is very complex. A simple limiting case is the following:
23.26. 'R'' $(a_2, \ldots a_n)$' for: '$\hat{x}_1\{(Ex_2, \ldots x_n) \cdot x_2 \varepsilon a_2 \cdot x_3 \varepsilon a_3 \ldots$ $\ldots x_n \varepsilon a_n \cdot R(x_1, \ldots x_n)\}$'. Cf. 19.21.

23.3. *Converses*

23.31. A relation with n terms has $n! - 1$ converses.
Explanation: $n! = 1 \cdot 2 \cdot \ldots n$; thus for $n = 3$, $n! = 1 \cdot 2 \cdot 3 = 6$; a triadic relation thus has $6 - 1 = 5$ converses, namely those holding between the following arguments: (1): 1, 3, 2; (2): 2, 1, 3; (3): 2, 3, 1; (4): 3, 1, 2; (5): 3, 2, 1.
23.32. '$R^{(a \ldots k \ldots u)}$' where '$a, k, u$' are variables for numbers between 1 and n, for: '$\hat{x}_a, \ldots \hat{x}_k, \ldots \hat{x}_u\{R(x_1, \ldots x_n)\}$'.
Example: '$R^{(231)}$' for: '$\hat{x}_2, \hat{x}_3, \hat{x}_1\{R(x_1, x_2, x_3)\}$'

23.4. *Domains and Fields*

23.41. 'D_1'R' for: '$\hat{x}_1\{(Ex_2, \ldots x_n)R(x_1, \ldots x_n)\}$'. Cf. 20.11
23.42. 'D_k'R' for: '$\hat{x}_k\{(Ex_1, \ldots x_{k-1}, x_{k+1}, \ldots x_n)R(x_1, \ldots x_k, \ldots x_n)\}$'
23.43. If R has n terms, 'C'R' for: 'D_1'$R \cup D_2$'$R \cup \ldots D_n$'R'
23.44. '$R \upharpoonright_k a$' for: '$\hat{x}_1, \ldots \hat{x}_n\{x_k \varepsilon a \cdot R(x_1, \ldots x_n)\}$'
23.45. '$R \lceil a$' for: '$\hat{x}_1, \ldots \hat{x}_n\{x_1, \ldots x_n \varepsilon a \cdot R(x_1, \ldots x_n)\}$'

23.5. *Partial Relations*

23.51. A relation with n terms contains $\binom{n}{m}$ partial relations with m terms.

Explanation: $\binom{n}{m} = \dfrac{n \cdot (n-1) \cdot (n-2), \ldots \{(n-(m-1)\}}{m!}$

It is the theorem used for calculating the coefficients of the binomial theorem or Pascal's triangle. For n and m from 1 to 10 one obtains the following table:

23.52. Number of partial relations:

$n =$	$m = 2$	3	4	5	6	7	8	9	10
2	1								
3	3	1							
4	6	4	1						
5	10	10	5	1					
6	15	20	15	6	1				
7	21	35	35	21	7	1			
8	28	56	70	56	28	8	1		
9	36	84	126	126	84	36	9	1	
10	45	120	210	252	210	120	45	10	1

Example: A relation with 4 terms $R(x, y, z, t)$ contains 6 partial dyadic relations (between $x - y$, $x - z$, $x - t$, $y - z$, $y - t$, and $z - t$), 4 triadic relations (between $x - y - z$, $x - y - t$, $x - z - t$, and $y - z - t$), and 1 tetradic relation (R). Moreover, each has its converse.

23.53. '$\left(R^{m}_{n}\right)$' for: 'the nth partial relation of m terms contained in R'.

Example: '$\left(R^{3}_{2}\right)$' for: 'the second triadic relation contained in R'.

LITERATURE: Carnap 1, 8.

V

VARIA

§ 24. CANONIC OR NORMAL FORM

Besides the method of evaluation given in § 4, there is another known as the 'canonic' or 'normal' form. Since it cannot be developed before the theory of rules (§ 9), the exposition of it has been postponed until now. Only a summary of the method is given here without any claim to rigor.

24.11. 'Canonic or normal form' for: 'a logical product of which each argument is a logical sum of variables or of negations of variables'. Example: '$(p \vee \sim q) \cdot (p \vee q) \cdot (\sim p \vee r)$' is a normal form.

24.12. Every sentence of system 8 can be transformed into an equivalent normal form in which each argument contains variables equiform with all the variables of the sentence. This transformation is accomplished by means of the rules which correspond (by the procedures of § 9) to the associative and distributive laws of the sum and the product (5.23–4, 5.53–4), the principle of double negation (5.12), the laws of De Morgan (5.27, 5.57), and laws 5.311 and 5.612.

Explanation: In practice this means that one must 'multiply' with '$\vee$' and '$\cdot$', as in algebra; substitute 'p' for '$\sim \sim p$', '$\sim p \cdot \sim q$' for '$\sim \cdot p \vee q$', '$\sim p \vee \sim q$' for '$\sim \cdot pq$', '$\sim p \vee q$' for '$p \supset q$', and '$p \supset q \cdot q \supset p$' for '$p \equiv q$'; and repeat these operations until the normal form is obtained. In this it is better to use the Peano-Russell notation with parentheses, since its similarities to algebra facilitate the 'multiplication'.

Example: Put in normal form the sentence '$(p \supset q) \supset (\sim q \supset \sim p)$'.

Applying 5.311 we obtain (1) $(\sim p \vee q) \supset (q \vee \sim p)$;
applying it again: (2) $\sim (\sim p \vee q) \vee (q \vee \sim p)$;
by De Morgan's law: (3) $(p \cdot \sim q) \vee (q \vee \sim p)$;
'multiplying' (4) $(p \vee q \vee \sim p)(\sim q \vee q \vee \sim p)$,

which is the normal form of the sentence.

24.13. *Rule:* the normal form is a law if and only if each argument of the product contains at least one variable with its equiform preceeded by a negation.

Explanation: In virtue of the rule founded on 6.13, '$p \vee \sim p$' is always true; on the other hand, by 6.26, if any argument of an alternative is true the whole is true; finally the product of true sentences is itself true (cf. 4.23). – Rule 24.13 thus enables us to evaluate a sentence.

LITERATURE: Hilbert A; Scholz 5; Quine 4; Łukasiewicz 7.

§ 25. MODAL LOGIC

This chapter contains the fundamental notions of a modal logic of sentences, i.e. of sentences stating necessary, possible, impossible, or contingent facts or on an alternative metalogical explanation, whose truth or falsity is necessary, possible, impossible or contingent.

25.1. *Monadic modal functors*

25.11. '*Lp*' or '□*p*' for: '*p* is necessary' or 'the sentence '*p* is necessarily true''.
Explanation: '*L*' (from 'logical' since what is logically true is necessarily true) is taken as the undefined functor in terms of which the remaining are defined.
25.12. '*Mp*' or '◇ *p*' for: '*NLNp*' or '∼ □ ∼ *p*'
Explanation: Read '*p* is possible'. '*M*' is from the German 'möglich'; '◇' was introduced by C. I. Lewis in 1918.
25.13. '*Up*' or '∼ ◇ *p*' for: '*LNp*'
Explanation: Read: '*p* is impossible'.
25.14. '*Zp*' or '∼ □ *p*' for: '*NLp*'
Explanation: Read: '*p* is contingent'. Although defined here in terms of not being necessary, the contingent functor is sometimes defined in terms of possibility: *KMpMNp*; or in another sense: '*p* is contingent' for '∼ *Lp* · ∼ *UP*'.
25.15. The 4 monadic functors *L, M, U, Z* give the 4 fundamental modalities; *L* and *M* are said to be positive, and *U* and *Z* negative modalities.

25.2. *Laws of Modals*

25.21. *CLpMp*
Explanation: This says that what is necessary is also possible. Taken as an axiom along with the definitions of 25.1 it yields the following laws:
25.22. *CUpZp*
25.23. *DLpUp*
25.24. *AMpZp*
25.25. *JLpZp*
25.26. *ELpNMNp*
25.27. *EMpNLNp*
25.28. *EUpLNp*
25.29. *EZpNLp*

25.3. *Dyadic modal functors*

25.31. '*C'pq*' or '*p* ⥽ *q*' for: '*LCpq*'
Explanation: This is usually read: '*p* strictly implies *q*', which is, however, a metalogical interpretation. Called the functor of 'strict implication', it was introduced by Lewis in 1918 to assimilate material implication to the every-day notion of implication and to show that the consequences of the paradoxes are unobjectionable.

25.32. '*E'pq*' for: '*KC'pqC'qp*'
Explanation: This gives us a 'strict equivalence'.

HISTORY: Modal logic was founded by Aristotle and further developed by his school and the Scholastics. In modern logic it was introduced by C. I. Lewis in 1918 with his system of 'strict implication', which posited 5 distinct systems S1 – S5. Since then still further systems have been developed.

LITERATURE: Lewis 1, Lewis L; Feys 3, Feys 4; Emch; Becker 1, Becker 2; Behmann 2; Łukasiewicz 8; Carnap 6, Carnap 7; Wright 1, Wright 2. – For history: Becker A; Bocheński 1, Bocheński 8.

§ 26. POLYVALENT LOGIC; COMBINATORY LOGIC; FORMALIZED METALOGIC

This chapter contains brief indications about three fields which logical research has recently developed. The fields have nothing more in common. Polyvalent logics are still being discussed. Combinatory logic is somewhat less discussed, partly because it is not so well known. Formalized metalogic, on the other hand, is a well established discipline.

26.1. *Polyvalent Logic*

By admitting only 2 values, a *bivalent* logic is formed; by admitting 3 values ('1', '½', '0', or '1', '2', '3'), a *trivalent* logic is obtained; in general, *n* admitted values (where '*n*' stands for any positive integer) gives a logic of *n* values. The number of *n*-adic truth functors in a logic with *m* values is m^{m^n}, from which we can form the following table:

values	2	3	4
monadic functors	4	27	256
dyadic functors	16	19, 683	4, 294, 967, 296

Thus in logics of more than two values it is possible to define many functors not translatable in terms of bivalent logic, e.g. the modal functors. On the other hand, certain laws of bivalent logic cease to be laws in trivalent or higher logics. For example, the principle of excluded middle does not hold in trivalent logic, since by substitution of $p/\frac{1}{2}$ and $q/\frac{1}{2}$ one obtains '$\frac{1}{2} \vee \frac{1}{2}$', which according to the definition of Łukasiewicz, gives '½'.

The principal functors of trivalent logic are defined, according to Łukasiewicz, as follows:

N	
1	0
½	½
½	1

A	1	½	0
1	1	1	1
½	1	½	½
0	1	½	0

C	1	½	0
1	1	½	0
½	1	1	½
0	1	1	1

K	1	½	0
1	1	½	0
½	½	½	0
0	0	0	0

26.2. *Combinatory Logic*

Combinatory logic is the theory of functors called 'combinators', which indicate a formal operation performed upon any type of expression. The principal combinators are the following: '*B*', called 'compositor', transforms an expression composed of 3 terms by grouping the 2nd and 3rd in a parenthesis. '*C*', called 'permutator', transforms an expression composed of 3 terms by inverting the order of the 2nd and 3rd terms. '*I*', called the 'identity functor', transforms a term into itself. 'W', called 'repetitor', transforms an expression composed of two terms by doubling the 2nd term.

Application of these combinators is not made directly to the expressions of classical mathematical logic, but to formulas in which the lambda ('λ') functor appears. This functor plays a role similar to that of variables with a circumflex accent ('$\hat{x}$') (15.111) but generalized. The expression '$\lambda a[M]$', for example, represents the operation which, applied to 'a', transforms it into the expression 'M'.

Combinators can be combined with one another, thereby giving rise to great possibilities for the simplification and generalization of logic.

26.3. *Formalized Metalogic*

Metalogic (cf. 2.16), also called 'semiotics', contains 3 parts: (1) *logical syntax:* the theory of the relations between signs; (2) *semantics:* the theory of the relations between signs and what they signify; (3) *pragmatics:* the theory of the relations between signs and their users. The first two parts have been formulated in conventional terms and formalized (7.51). This procedure has the following advantages: it enables us to make a more exact analysis of logical notions, to axiomatize metalogic, to make a rigorous study of systems with respect to completeness and non-contradiction (7.6) and the independence of axioms (7.7), and to define with precision the metalogical terms that are constantly used in logic, such as 'system', 'deduction', 'term', 'variable', 'expression', etc. Systems are then treated as classes.

The technical procedure consists generally of giving a double translation to the terms of the system being examined, as was done in § 9. With each term a metalogical sign is coordinated, and the fact that term *Y* follows term *X* is expressed by a special symbol.

Metalogic has shown itself to be fruitful in philosophical conclusions, notably with respect to the definition of such terms as 'truth', 'signification', etc.

HISTORY: Polyvalent logic was discovered independently by Łukasiewicz (1920) and Post (1921). It was cultivated particularly in Poland, where Wajsberg gave the first axiomatization of it (1931). It has been discussed more recently in relation to the calculus of probabilities and to its philosophical implications. Reichenbach has developed it for quantum mechanics. – Combinatory logic, which is the most recent branch of our science, was begun by Schönfinkel in 1924 and developed by Curry and Kleene; Church in 1936 provided a synthesis of previous work. – Metalogic (called also 'semantics') has precursors in the theory of supposition and the treatises 'de modis significandi' of mediaeval logic. But as an exact discipline its history is very recent. The 'metamathematics' of Hilbert (1905), the speculations of the Vienna circle on language (1929 ff.), the rigorous axiomatization of the Polish system are the three main sources of the new discipline. Its development is due primarily to the work of Carnap, Gödel, Leśniewski, and Tarski.

LITERATURE: § 26.1: Post; Łukasiewicz 1, Łukasiewicz 4; Feys 3; Wajsberg; Hempel; Rosser T; Reichenbach 2.
§ 26.2: Schönfinkel; Curry 1, 4; Church 4; Feys 7.
§ 26.3: Carnap 3, Carnap 4, 7; Tarski 4; a good resumé in Quine 3; Schröter 1; Scholz 2, Scholz 6; Church 6.

§ 27. THE SYNTACTICAL CATEGORIES

This chapter provides a further example of metalogic. The syntactical categories, mentioned in 1.22, which are of great importance for philosophy, are defined and their various kinds enumerated.

27.1. *Definitions*

27.11. '$SS(x, y, l)$' for: '$(u, v) :: Fl(u, l) :\cdot \supset :\cdot P(x, u, l) \cdot$
$\cdot Sb(y, x, u, v) \cdot \vee \cdot P(y, u, l) \cdot Sb(x, y, u, v) : \supset : Fl(v, l)$'
Explanation: $SS(x, y, l)$ is the triadic relation mentioned in 1.22 whereby x and y can be mutually substituted for one another in the language l; this is exactly defined in what follows. '$Fl(u, l)$' says that u is a formula or expression (1.11) in the language l; '$P(x, u, l)$' that x is a part of u in the language l, i.e. either x is a sign in the language l or one of a series of signs in l; '$Sb(y, x, u, v)$' says that v is a substitution of y for x in u.
27.12. '$SC(al)$' for: '$(x, y) \cdot x, y, \varepsilon\, a \supset SS(x, y, l)$'
Explanation: 'SC' for 'syntactical category'; but what is here defined is not this, but the relation 'a is an SC of the language l', i.e. the case in which all elements of a can be mutually substituted for one another in the language. SC generically is the class of all classes a such that for some l we have $SC(al)$, i.e. it is the domain of SC.

27.2. *Division of SC*

27.21. '*ESC*' for: 'Elementary or fundamental *SC*, which appears only as an argument and never as a functor'.
Explanation: Signs belonging to *ESC* mean something which may have a property but cannot be a property.
The most common *ESC* are:
27.211. '*n*' for: 'individual name' (cf. 1.33)
27.212. '*s*' for: 'sentence' (cf. 1.31)
27.213. '*u*' for: 'universal or class name' (cf. 15.11)
27.22. '*FSC*' for: 'Functional *SC* whose elements appear as both functors and arguments'.
Explanation: The elements of *FSC* are called 'functors' (cf. 1.34). They may be classified according to 3 criteria:
27.221. According to the *SC* of their arguments, we can distinguish: *name-determining* functors, *sentence-determining* functors, and *universal* or *class-determining* functors.

Examples: The predicates φ, ψ, χ are name-determining functors; $\vee$, $\supset$, $\sim$, are sentence-determining functors; $-$, $\cup$, $\cap$ are class-determining functors.

27.222. According to the number of their arguments, we can distinguish: *monadic*, *dyadic*, *triadic*, ... *n-adic* functors (cf. 1.45)

Examples: $\sim$ is monadic, $\vee$ is dyadic, *SS* in 27.11 is triadic, *Sb* in 27.11 is tetradic.

27.223. According to the *SC* of the whole formula resulting from the functor and its arguments, we can distinguish:

name-forming, *sentence-forming*, and *universal* or *class-forming* functors.

Examples: The description functor (14.22) is a name-forming, monadic, sentence-determining functor; the relation *R* in 18.14 is a sentence-forming, dyadic, name-determining functor; the antecedent-class description in 19.12 $\vec{R}$'*y* is a class-forming, monadic, sentence-determining functor.

27.23. *The Ajdukiewicz method for determining the SC of a functor:*

Form a fraction, the numerator of which represents the *SC* of the formula which it forms, and the denominator the *SC* of the arguments which it determines; if more than one argument is determined, the letters representing each are written in the denominator separated by commas.

Examples: The examples of 27.223 would be presented, according to this method, as follows: $\frac{\mathrm{n}}{\mathrm{s}}$; $\frac{\mathrm{s}}{\mathrm{n,n}}$; $\frac{\mathrm{u}}{\mathrm{s}}$.

For *CNpCpNq* we would have

$$\frac{s}{\frac{\mathrm{s}}{\mathrm{s}},\ \frac{\mathrm{s}}{\mathrm{s}\ ,\ \frac{\mathrm{s}}{\mathrm{s}}}}$$

In this the '*s*' in the numerator indicates that the whole expression that is formed is a sentence; the first fraction in the denominator represents '*Np*', a monadic, sentence-determining, sentence-forming functor, while the second fraction represents '*CpNq*', which is itself a sentence formed by determining a sentence '*p*' and another sentence '*Nq*' formed out of another sentence '*q*' and a monadic, sentence-determining functor '*N*'.

27.3. *Fundamental law of SC*

27.31. $(x, l) : Fl(x, l) \cdot \supset \cdot (Ea) \cdot SC(a, l) \cdot x \, \varepsilon \, a$

Explanation: All formulas or expressions of a language belong to a *SC* of that language.

HISTORY: The idea of SC comes from Husserl, although something like it is found in Aristotle and the Scholastics. The rigorous development of the theory is due to Leśniewski and Ajdukiewicz.

LITERATURE: Leśniewski 2, Ajdukiewicz 2; Bocheński 7.

TABLE OF LOGICAL SIGNS

2.13. ⌜ ⌝
2.3. ( ), [ ], { }
2.4. ·, :, ·:, : :
3.11. *1, 0*
3.12. =
3.23. —, ~, *N*
3.32. *V*, *A*, *B*, *C*, *D*, *E*, *F*, *G*, *O*, *X*, *M*, *L*, *K*, *J*, *I*, *H*
3.41. ∨, *A*
3.51. ⊃, →, *C*
3.61. |, *D*
3.71. ·, &, *K*
3.81. ≡, ~, *E*
8.11. *p*, *q*, *r*, *s*
8.51. /
9.5. —, +, →, =, /, ×, ⊢, $\frac{\vdash X}{\vdash Y}$, ∞
10.001. *a*, *b*, *m*, *A*, *E*, *I*, *O*
11.11. *a*, *b*, *c*, *d*
11.12. *x*, *y*, *z*, *t*
11.13. *φ*, *ψ*, *χ*
11.15. *φx*
11.21. (*x*), *Πx*,
11.22. (*Ex*), (∃*x*), *Σx*
13.11. *φ*(*x*, *y*)
13.12. (*x*, *y*), *Πxy*
13.13. (*Ex*, *y*), *Σxy*
14.11. =
14.12. ≠
14.16. *I*
14.17. *J*
14.22. (℩*x*)
14.23. *E!*
15.11. *x̂*
15.111. *λ*
15.12. *Cls*, *α*, *β*, *γ*
15.13. *ε*
15.14. ~ *ε*
15.21. — *α*
15.22. ∪
15.23. ∩
15.24. ||
15.25. ⊂
15.26. =
15.41. **V**
15.42. Λ
15.43. ∃!
15.61. [*x*]
15.62. [*x*, *y*]
15.63. 1
15.64. 2
18.11. *x̂ŷ*{*φx*, *y*}
18.12. *Rel*
18.14. *xRy*
18.21. ∸
18.22. ∪̇
18.23. ∩̇
18.24. ||̇
18.25. ⊂̇
18.26. ≐
18.27. **V̇**
18.28. Λ̇
18.29. ∃̇!
19.11. *R*'*y*
19.12. $\overrightarrow{R}$'*y*
19.13. $\overleftarrow{R}$'*x*
19.14. *sg*'*R*
19.15. *gs*'*R*
19.21. *R*''*β*
19.31. *Ř*
19.32. *Cnv*
20.11. *D*'*R*
20.12. *ᗡ*'*R*
20.13. *C*'*R*
20.31. *α* ⌉ *R*
20.32. *R* ⌈ *β*
20.34. *R* ⌈ *α*
20.35. ↑
20.37. ↓
20.41. *1* → *Cls*
20.42. *Cls* → *1*
20.43. *1* → *1*
21.11. *R* | *S*
21.12. R^2
21.13. R^3

21.14. R^n
21.15. R^0
21.21. *her*
21.22. R_*
21.23. R_{po}
21.31. B
21.32. Min_R
21.33. Max_R
21.41. $R \dagger S$
21.42. $\overline{smor}$
21.43. *Smor*
22.11. *refl*
22.12. *irr*
22.21. *sym*
22.22. *as*
22.31. *trans*
22.32. *intr*
22.41. *sim*
22.42. *aeq*
22.51. *connex*
22.52. *ser*
23.12. Rel_n
23.21. $R_1'(x_2, \ldots x_n)$
23.22. R_k'
23.23. $\vec{R}_1'$
23.24. $\vec{R}_k'$
23.25. $sg_k'R$
23.26. $R''(a_2, \ldots a_n)$
23.32. $R^{(a \ldots k \ldots u)}$
23.41. $D_1'R$
23.42. $D_k'R$
23.43. $C'R$
23.44. $R \upharpoonright_k a$
23.45. $R \upharpoonleft a$
23.53. (R^m_n)
25.11. L, $\Box$
25.12. M, $\Diamond$
25.13. U
25.14. Z
25.31. C', $\dashv$
25.32. E'
26.2. λ,
B, C, I, W
27.11. *SS*
27.12. *SC*
27.21. *ESC*
27.211. n
27.212. s
27.213. u
27.22. *FSC*

BIBLIOGRAPHY

Abbreviations: 'JSL' 'The Journal of Symbolic Logic'.
'PM' 'Principia Mathematica'.

ACKERMANN, W. (1) *Ein System der typenfreien Logik I*, Leipzig, 1941.

ACKERMANN, W. (2) *Solvable Cases of the Decision Problem*, Amsterdam, 1955. Cf. Hilbert

AJDUKIEWICZ, K. (1) *Założenia logiki tradycyjnej*, (Presuppositions of trad. Logic), Przegl. Filozoficzny 29, 1926/7.

AJDUKIEWICZ, K. (2) *Die syntaktische Konnexität*, Studia Philosophica, (Lwów) I, 1935.

AJDUKIEWICZ, K. (3) *Über die Anwendbarkeit der reinen Logik auf philosophische Probleme*. Actes du VIe Congr. Int. de Philos. Prague, 1936.

ARISTOTLE, *Prior and Posterior Analytics*, ed. D. Ross, Oxford, 1949.

BANKS, P. *On the philosophical interpretation of Logic: an Aristotelian Dialogue*, Dominican Studies, (Oxford) III/2 (1950), p. 139 ff.

BECKER, A. *Die Aristotelische Theorie der Möglichkeitsschlüsse*, Berlin, 1933.

BECKER, O. (1) *Zur Logik der Modalitäten*, Jahrb. f. Phil. u. Phän. Forsch. 11 (1930).

BECKER, O. (2) *Untersuchungen über den Modalkalkül*, Meisenheim, 1952.

BEHMANN, H. (1) *Zu den Widerspruchen der Logik und Mengenlehre*, Jahresber. d. Math. Ver. 40 (1931).

BEHMANN, H. (2) *Die typenfreie Logik und die Modalität*, Actes du XIème Congr. d. Philos. Bruxelles, XIV (1953), p. 88 ff.

BENNET, A. A. and CH. A. BAYLIS, *Formal Logic*, New York, 1939.

BERKELEY, E. C. *Conditions affecting the application of symbolic logic*, JSL 7, 1942.

BERNAYS, P. *A System of axiomatic Set Theory*, JSL 2, 1937; 6, 1941; 7, 1942; 8, 1943.

BETH, E. W. (1) *Inleiding tot de wijsbegeerte der wiskunde*, Antwerp, 1940; 2nd edit. 1948.

BETH, E. W. (2) *Summulae Logicales*, Groningen, 1942.

BETH, E. W. (3) *Geschiedenis der Logica*, Den Haag, 1944.

BETH, E. W. (4) *Symbolische Logik und Grundlegung der exakten Wissenschaften*. (Bibl. Einf. i. d. Stud. d. Philos. ed. I. M. Bocheński, N. 3) Bern, 1948.

BETH, E. W. (5) *Les fondements de mathématique*, Louvain, 1950.

BLACK, H. (1) *The Nature of Mathematics*, New York 1934, repr. 1952.

BLACK, H. (2) *A New Method of Presentation of the Theory of the Syllogism*, Journal of Philosophy, 1945.

BOCHEŃSKI, J. M. (1) *Notes historiques sur les propositions modales*, Révue des Sciences Philos et Théol. 26. 1937.

BOCHEŃSKI, J. M. (2) *De consequentiis Scholasticorum earumque origine*, Angelicum 15, 1938.

BOCHEŃSKI, J. M. (3) *La logique de Théophraste*, Fribourg, 1947.

BOCHEŃSKI, J. M. (4) *On the Categorical Syllogism*, Dominican Studies (Oxford) I 1948, p. 35 ff.

BOCHEŃSKI, J. M. (5) *On Analogy*. The Thomist XI/4 (1948) p. 424 ff.

BOCHEŃSKI, J. M. (6) *On Syntactical Categories.* New Scholasticism 23/5 (1949), p. 257 ff.

BOCHEŃSKI, J. M. (7) *Ancient Formal Logic*, Amsterdam, 1951.

BOCHEŃSKI, J. M. (8) *Formale Logik: Problemgeschichte*, Freiburg-i-B, 1956, Eng. trans. by Ivo Thomas, Notre Dame, 1960.

BOEHNER, P. *Medieval Logic. An Outline of Its Development from 1250-c.1400*, Manchester, 1952.

BOOLE, G. (1) *The Mathematical Analysis of Logic*, Cambridge, 1847, repr. 1947.

BOOLE, G. (2) *An Investigation of the Laws of Thought*, London, 1854, repr. New York, n.d. *Cf. Celebration of the centenary of the LAWS OF THOUGHT*, Royal Irish Acad. 57/6, 1955.

CARNAP, R. (1) *Abriss der Logistik*, Wien, 1929.

CARNAP, R. (2) *Logische Syntax der Sprache*, Wien, 1934, trans. *The Logical Syntax of Language*, New York, 1937.

CARNAP, R. (3) *Foundations of Logic and Mathematics.* Intern. Encycl. of Unified Science I/3, Chicago, 1939.

CARNAP, R. (4) *Introduction to Semantics*, Cambridge Mass. 1942 3rd edit. 1948.

CARNAP, R. (5) *Formalization of Logic*, Cambridge Mass. 1943.

CARNAP, R. (6) *Modalities and quantification.* JSL 11, 1946.

CARNAP, R. (7) *Meaning and Necessity: A Study in Semantics and Modal Logic*, Chicago 1947; 2nd edit. 1956.

CARNAP, R. (8) *Introduction to Symbolical Logic and its Applications*, New York, 1958.

CHURCH, A. (1) *A Bibliography of Symbolic Logic*, JSL 1, 1936, 3, 1938.

CHURCH, A. (2) *A Formulation of the Simple Theory of Types*, JSL 5, 1940.

CHURCH, A. (3) *Conditioned Disjunction as a primitive Connective for the Propositional Calculus.* Portugaliae Math 7/2 (1948) p. 87 ff.

CHURCH, A. (4) *The Calculus of Lambda Conversion.* Princeton, 1941.

CHURCH, A. (5) *A Brief Bibliography of Formal Logic*, Proc. Amer. Acad. Arts and Sci, 80, 1952.

CHURCH, A. (6) *Introduction to Mathematical Logic*, Vol I, Princeton, 1956.

CHWISTEK, L. (1) *Antynomie Logiki formalnej.* Przegl. Filozoficzna 24, 1921.

CHWISTEK, L. (2) *The Theory of Constructive Types.* Ann. de la soc. Pol. des mathèm. 2, 1924.

CHWISTEK, L. (3) *New Foundations of formal Metamathematics*, JSL 3, 1938.

COOLEY, J. C. *A Primer of Formal Logic*, New York, 1942.

COPI, I. *Symbolical Logic*, New York, 1954.

COUTURAT, L. *La logique de Leibniz d'après des documents inédits.* Paris, 1901.

CURRY, H.B. (1) *Grundlagen der kombinatorischen Logik.* Amer. Journ. of Math. 52, 1930.

CURRY, H.B. (2) *On the Use of Dots as Brackets in Logical Expressions*, JSL 2, 1937.

CURRY, H.B. (3) *A Mathematical Treatment of the Rules of the Syllogism*, Mind, 45 (1936) p. 209 ff.

CURRY, H.B. (4) R. FEYS, W. CRAIG, *Combinatory Logic*, Amsterdam, 1958.

DOPP, J. Leçons de logique formelle, 3 vols. Louvain, 1950.

DUBLISLAV, W. (1) *Die Definition*, 2nd edit. Berlin 1927.

DUBLISLAV, W. (2) *Die Philosophie der Mathematik in der Gegenwart*, Berlin, 1932.

DÜRR, K. (1) *Aussagenlogik im Mittelalter*, Erkenntnis 7, 1938.

DÜRR, K. (2) *Lehrbuch der Logistik*, Basel, 1954.

EMCH, A. F. *Deducibility with respect to Necessary and Impossible Propositions*, JSL 2, 1937.

FEYS, R. (1) *La transcription logistique du raisonnement*, Rev. Néoscol. d. Philos. 26–27, 1924–25.

FEYS, R. (2) *La raisonnement en termes de faits dans la logique russellienne*, ibid. 29–30, 1927–28.

FEYS, R. (3) *Les logiques nouvelles de la modalité*, ibid. 40–41, 1937–38.

FEYS, R. (4) *Directions nouvelles de la logistique aux États-Unis*, ibid. 44, 1946.

FEYS, R. (5) *Logistiek I.* Antwerpen-Nijmegen, 1944.

FEYS, R. (6) *Les méthodes récentes de déduction naturelle*, Rev. philos. de Louvain 44, 1946.

FEYS, R. (7) *La technique de la logique combinatoire*, ibid. 44, 1946. Cf. Curry 4.

FITCH, F. B. (1) *A System of Formal Logic without an analogue to the Curry W-Operator*, JSL 1, 1936.

FITCH, F. B. (2) *Symbolic Logic*, New York, 1952.

FRAENKEL, A. et Y. BAR-HILLEL, *Le problème des antinomies et ses développements récentes*, Rev. d. metaphys et d. morale, 46, 1939.

FREGE, G. (1) *Begriffsschrift*, Halle 1879.

FREGE, G. (2) *Grundlagen der Arithmetik*, Breslau, 1884; *Foundations of Arithmetic*, trans. by J. Austin, Oxford, 1953.

FREGE, G. (3) *Funktion und Begriff*, Jena, 1891, trans. in *Philosophical Writings of Gottlob Frege*, by P. Geach, M. Black, Oxford, 1952.

FREGE, G. (4) *Grundgesetze der Arithmetik*, Jena I 1893, II 1903.

GENTZEN, G. *Untersuchungen über das logische Schliessen*, Math. Zeitschr. 39, 1934.

GÖDEL, K. *Über formal unentscheidbare Sätze der Principia Mathematica und verwandter Systeme*, Monatshefte Math. Phys. 38, 1930.

GONSETH, F. (1) *Les fondements des mathématiques*, Paris, 1926.

GONSETH, F. (2) *Qu'est-ce que la logique*, Paris, 1937.

GONSETH, F. (3) *Les entretiens de Zürich sur les fondements et la méthode des sciences mathématiques*, Zürich, 1941.

GOODSTEIN, R. L. *Mathematical Logic*, Leicester, 1957.

GREENWOOD, TH. *Les fondements de la logique symbolique*, 2 vols., Paris, 1938.

HEMPEL, C. G. *A purely topological form of non-Aristotelian Logic*, JSL 2, 1937.

HEYTING, A. (1) *Die formalen Regeln der intuitionistischen Logik*, Sitzungsber. d. Preuss. Ak. d. Wiss. Math. Phys. Kl. 1930.

HEYTING, A. (2) *Die intuitionistische Grundlegung der Mathematik*, Erkenntnis 2, 1932.

HEYTING, A. (3) *Mathematische Grundlagenforschung: Intuitionismus – Beweistheorie*. Erg. d. Math. III. 4, Berlin, 1934.

HEYTING, A. (4) *Intuitionism: An Introduction*, Amsterdam, 1956.

HILBERT, D. und W. ACKERMANN, *Grundzüge der theoretischen Logik*, Berlin, 1928, trans. *Principles of Mathematical Logic*, New York, 1950.

HILBERT, D. und P. BERNAYS, *Grundlagen der Mathematik*, Berlin, I, 1934; II, 1939, repr. Ann Arbor, Mich. 1944.

HUNTINGTON, E. V. *Sets of Independent Postulates for the Algebra of Logic*, Trans. Amer. Math. Soc. 5, 1904.

HUSSERL, E. *Logische Untersuchungen*, Halle, I, 1900; II, 1901.

JAŚKOWSKI, S. *On the rules of Suppositions in Formal Logic*, Studia Logica 1, Warsaw, 1934.

JÖRGENSEN, J. (1) *A Treatise of Formal Logic*, 3 vols. London-Copenhagen, 1931.

JÖRGENSEN, J. (2) *Einige Hauptpunkte der Entwicklung der formalen Logik seit Boole*, Erkenntnis 5, 1935.

JORDAN, Z. *The Development of Mathematical Logic and of Logical Positivism in Poland*, London, 1945.
KEYNES, J. N. *Studies and Exercises in Formal Logic*, London 1894; 4th edit. 1906.
KLEENE, S. C. (1) *A theory of positive integers in formal logic*, Am. Journ. of Math. 57, 1935.
KLEENE, S. C. (2) *Introduction to Metamathematics*, New York, 1952.
KLUG, U. *Juristische Logik*, Berlin, 1951.
KURATOWSKI, K. *Sur la notion de l'ordre dans la théorie des ensembles*, Fundamenta Math. 2, 1921.
LADRIÈRE, J. *Les limitations internes des formalismes*, Louvain, 1957.
LEBLANC, H. *An Introduction to Deductive Logic*, New York, 1955.
LEŚNIEWSKI, S. (1) *Über die Grundlagen des Ontologie*, C. r. de la Soc. des Sc. et d. Lettres de Varsovie, 61, III, 1930.
LEŚNIEWSKI, S. (2) *Grundzüge eines neuen Systems der Grundlagen der Mathematik*, Warschau, 1938.
LEŚNIEWSKI, S. (3) *Einleitende Bemerkungen zur Fortsetzung meiner Mitteilung u. d. T. 'Grundzüge...'*, Warschau, 1938.
LEWIS, C. I. (1) *A Survey of Symbolic Logic*, Berkeley, 1918.
LEWIS, C. I. (2) and C. H. LANGFORD, *Symbolic Logic*, New York 1936, repr. 1952.
ŁUKASIEWICZ, J. (1) *O Logice tròjwartościowej* (On three-valued logic), Ruch Filozoficzny 5, 1920.
ŁUKASIEWICZ, J. (2) *Logika dwuwartosciowa* (Two-valued logic), Przeglad Filozoficzny 23, 1931.
ŁUKASIEWICZ, J. (3) *Elementy logiki matematycznej*, Warszawa, 1929.
ŁUKASIEWICZ, J. (4) *Philosophische Bemerkungen zu mehrwertigen Systemen des Aussagenkalküls*, C.r. Soc. d. Sc. et d. Lettres Varsovie, C. III, 23, 1930.
ŁUKASIEWICZ, J. (5) *Zur Geschichte der Aussagenlogik*, Erkenntnis 5, 1935–36.
ŁUKASIEWICZ, J. (6) *Die Logik und das Grundlagenproblem*, in Gonseth 3.
ŁUKASIEWICZ, J. (7) *Aristotle's Syllogistic from the Standpoint of Modern Formal Logic*, Oxford, 1951; 2nd edit. 1955.
ŁUKASIEWICZ, J. (8) *A system of Modal Logic*, Actes du XIème Congr. Int. d.Philos. Bruxelles, 1953, Vol. XIV, p. 82, ff.
ŁUKASIEWICZ, J. (9) *A System of Modal Logic*, in Journ. of Computing Systems I, 1953.
ŁUKASIEWICZ, J. and A. TARSKI, *Untersuchungen über den Aussagenkalkül*, C. R. Soc. d. Sc. et d. Lettres Varsovie, Cl. III, 23, 1930.
MATES, B. *Stoic Logic* (Diss), Berkeley, Los Angeles, 1953.
MENGER, K. *Moral, Wille, Weltgestaltung, Grundlegung der Logik der Sitten*, Wien, 1934.
MENNE, A. *Logik und Existenz*, Meisenheim, 1954.
MILLER, J. W. *The Structure of Aritotelian Logic*, London, 1938.
MOISIL, G. C. (1) *Recherches sur l'algèbre de la logique*, Ann. scient. de l'Université de Jassy 22, 1936.
MOISIL, G. C. (2) *Recherches sur le syllogisme*, ibid. 25, 1939.
MOODY, E. *Truth and Consequence in Mediaeval Logic*, Amsterdam, 1953.
MOORE, G. E. *Russell's 'Theory of Descriptions'* in *The Philosophy of Bertrand Russell*, edit. P. A. Schilpp, Evanston, Chicago, 1944.
MORGAN, A. DE (1) *Formal Logic*, London 1847; repr. 1926.
MORGAN, A. DE (2) *On the Syllogism No. IV and on the Logic of Relations*, Trans. Cambr. Philos. Soc. 10, 1864.

MOSTOWSKI, A. *Sentences Undecidable in Formalized Arithmetic*, Amsterdam, 1952.
NICOD, J. A. *A reduction in the number of primitive propositions of logic*, Proc. Cambr. Philos. Soc. 19, 1917/20.
OPPENHEIM, F. E. *Outline of a logical Analysis of Law*, Philos. of Science 11, 1944.
PEANO, G. (1) *Formulaire mathématique*, I Turin, 1895; II/1 Turin, 1897; II/2 Turin 1898; II/3 Turin 1899; III Paris, 1901; IV Turin, 1902–03.
PEANO, G. (2) *Formulario matematico*, V Torino, 1905–08.
PEIRCE, C. S. *Collected Papers*, Vols. 1–6 edit. C. Hartshorne and P. Weiss; Vols. 7–8 edit. A. Burks, Cambridge, 1931–36; 1958.
P.M., A. N. WHITEHEAD and B. RUSSELL, *Principia Mathematica*, Cambridge, I, 1910, II, 1912, III, 1913; 2nd edit. 1925–27; repr. 1950.
POPPER, K.R. *New Foundations for Logic*, Mind, 56, 1947.
POST, E. L. *Introduction to a general theory of propositions*, Am. Journ. of Math, 43, 1921.
PRIOR, A. N. *Formal Logic*, Oxford, 1955.
QUINE, W. V. O. (1) *A System of Logistic*, Cambridge, Mass. 1934.
QUINE, W. V. O. (2) *On the Theory of Types*, JSL 3, 1938.
QUINE, W. V. O. (3) *Mathematical Logic*, New York, 1940; 2nd edit. Cambridge, 1951.
QUINE, W. V. O. (4) *Elementary Logic*, Boston, 1941.
QUINE, W. V. O. (5) *On the Logic of Quantification*, JSL 10, 1945.
QUINE, W. V. O. (6) *New Foundations of Mathematical Logic*, Am. Math. Monthly 44, 1937.
QUINE, W. V. O. (7) *Three grades of Modal Involvement*, Actes du XIème Congr. Int. Philos. Bruxelles, 1953, Vol. XIV, p. 65 ff.
QUINE, W. V. O. (8) *Methods of Logic*, New York, 1950.
QUINE, W. V. O. (9) *From a Logical Point of View*, Cambridge, 1953.
RAMSEY, F. P. *The Foundations of Mathematics and Other Essays*, London, 1931; repr. 1954.
REICHENBACH, H. (1) *Elements of Symbolic Logic*, New York, 1947.
REICHENBACH, H. (2) *Philosophische Grundlagen der Quantenmechanik*, Zürich, 1949.
ROSENBLOOM, P. C. *The Elements of Mathematical Logic*, New York, 1950.
ROSSER, J. B. *Logic for Mathematicians*, New York, 1953.
ROSSER, J. B. and A. R. TURQUETTE, *Many-valued Logics*, Amsterdam, 1952.
RUSSELL, B. A. W. (1) *The Principles of Mathematics*, Cambridge, 1903; 2nd edit. 1938; repr. 1951.
RUSSELL, B. A. W. (2) *On Denoting*, Mind 14, 1905; repr. in *Logic and Knowledge; Essays* 1901–1950, New York, 1956.
RUSSELL, B. A. W. (3) *Introduction to Mathematical Philosophy*, London, 1919.
RUSSELL, B. A. W. (4) *Mathematical Logic as based on the Theory of Types*, Am. Journ. of Math. 3, 1908, p. 222 ff.
Cf. P.M.
RÜSTOW, A. *Der Lügner, Theorie, Geschichte und Auflösung*, (Diss. Erlangen), Leipzig, 1910.
SALAMUCHA, J. *Pojawienie sie zagadnien antynominalnych na gruncie Logiki sredniowiecznej* (Appearance of the problem of paradoxes in mediaeval logic), Przeglad Filozoficzny, 40, 1937.
SCHMIDT, A. *Mathematische Grundlagenforschung*, Enz. d. math. Wiss. Bd. I, 1, Heft 1, Teil II.
SCHÖNFINKEL, M. *Über die Bausteine der mathematischen Logik*, Math. Ann. 42, 1925.

SCHOLZ, H. (1) *Geschichte der Logik*, Berlin, 1931.

SCHOLZ, H. (2) *Leibniz und die mathematische Grundlagenforschung*, Jahresber. d. deutsch. Math. Vers. 52, 1942.

SCHOLZ, H. (3) *Metaphysik als strenge Wissenschaft*, Köln, 1941.

SCHOLZ, H. (4) *Logik, Grammatik, Metaphysik*, Arch. f. Rechts- u. Soz.-Philos. XXXVI/3, p. 393 ff.

SCHOLZ, H. (5) *Grundzüge der mathematischen Logik*. 2 Bde. Münster, 1950–51.

SCHOLZ, H. (6) *Zur Erhellung des Verstehens*, in Geistige Gestalten und Probleme. Eduard Spranger zum 60 Geburtstag. Leipzig, 1942, p. 291.

SCHOLZ, H. und H. HERMES, *Mathematische Logik*, Enz. d. math. Wiss. Bd. I 1, Heft 1, Teil 1.

SCHRÖDER, E. *Vorlesungen über die Algebra der Logik*, Leipzig I, 1890; II 1, 1891; II 2, 1905; III, 1895.

SCHRÖTER, K. (1) *Ein allgemeiner Kalkülbegriff*, Leipzig, 1941.

SCHRÖTER, K. (2) *Axiomatisierung der Fregeschen Aussagekalküle*, Leipzig, 1943.

SKOLEM, T. *Über einige Grundlagenfragen der Mathematik*, Skrifter utgitt av Det Norske Videnskaps-Akademi, Oslo, 1929.

SOBOCINSKI, B. (1) *Aksjomatyzacja implikacyjno-Konjunkcyjnej teorii dedukcji*, Przegl. Filozoficzny, 38, 1935.

SOBOCINSKI, B. (2) *Aksjomatyzacja pewnych wielowartościowych systemów teorii deduckji* (Axiomatization of some polyvalent systems of the theory of deduction) Rocznicki Zrzesz. asyst. Uniw. J. P. Warszawa, 1936.

SOBOCINSKI, B. (3) *An Investigation of Protothetic*, Edit. de l'Instit. d'Etudes Polon. en Belgique, Brussels, 1949.

SUPPES, P. *Introduction to Logic*, New York, 1957.

TARSKI, A. (1) *Fundamentale Begriffe der Methodologie der deduktiven Wissenschaften*, Monatshefte, f. Math. u. Phys. 37, 1930.

TARSKI, A. (2) *Der Wahrheitsbegriff in den formalisierten Sprachen*, Studia Philos. (Lwów) I, 1935.

TARSKI, A. (3) *Wahrscheinlichkeitslehre und mehrwertige Logik*, Erkenntnis 5, 1936.

TARSKI, A. (4) *Grundzüge des Systemenkalküls* I, Fundam. Math. 25–26, 1935–36.

TARSKI, A. (5) *On the Calculus of Relations*, JSL 6, 1941.

TARSKI, A. (6) *Introduction to Logic and to Methodology of deductive Sciences*, New York, 1941.

TARSKI, A. (7) *Logic, Semantics, Metamathematics. Papers from 1923 to 1938*, trans. J. H. Woodger, Oxford, 1956.

TARSKI, A. (8) and R. ROBINSON, *Undecidable Theories*, Amsterdam, 1953.

THOMAS, Ivo (1) *Logic and Theology*, Dominican Studies I (1948).

THOMAS, Ivo (2) *CS(n): An extension of CS*, Dominican Studies II (1949).

THOMAS, Ivo (3) *A New Decision Procedure for Aristotle's Syllogistic*, Mind, 61, 1952.

THOMAS, Ivo (4) *The Faris System and Syllogistic*, JSL 20, 1955.

TURING, A. M. *The Use of Dots as Brackets in Church's System*, JSL 7, 1942.

USHENKO, A. M. *The Problems of Logic*, London, 1941.

WAJSBERG, M. *Aksjomatyzacja trojwartsciowego rachunku zdań* (Axiomatization of the three-valued calculus of sentences), C. r. Soc. Sc. et Lettres de Varsovie, 24, 1931.

WEDBERG, A. *The Aristotelian Theory of Classes*, Ajatus, Eripainos Filosofisen Yhdistuksa, Vuosikirjada (Helsinki) V (1948), p. 299 ff.

WHITEHEAD, Cf. PM.

WIENER, N. *A Simplification of the Logic of Relations*, Proc. Cambr. Philos. Soc. 17, 1912/14.

WITTGENSTEIN, L. *Tractatus Logico-Philosophicus*, London, 1922, repr. London 1951.

WRIGHT, G. H. VAN (1) *An Essay in Modal Logic*, Amsterdam, 1951.

WRIGHT, G. H. VAN (2) *A New System of Modal Logic*, Actes du XIème Congr. Int. de Philos. Bruxelles, 1953, Vol. V, p. 59 ff.

WOODGER, J. H. (1) *The Axiomatic Method in Biology* (with appendix by Tarski), Cambridge, 1937.

WOODGER, J. H. (2) *The Formulation of a Psychological Theory*, Erkenntnis 7, 1937.

WOODGER, J. H. (3) *The Technique of Theory Construction*, Intern. Encyc. Unified Sci. II/5, Chicago, 1935.

WOODGER, J. H. (4) *Biology and Language*, Cambridge, 1952.

ZERMELO, E. *Untersuchungen über die Grundlagen der Mengenlehre*, Math. Ann. 65, 1908.